THE PILTDOWN FORGERY

The Piltdown Forgery

¤ ¤ ¤

J. S. Weiner

Fiftieth anniversary edition,
with a new introduction and afterword by

Chris Stringer

OXFORD
UNIVERSITY PRESS

OXFORD
UNIVERSITY PRESS

Great Claredon Street, Oxford OX2 6DP

Oxford University Press is a department of the University of Oxford.
It furthers the University's objective of excellence in research, scholarship,
and education by publishing worldwide in

Oxford New York

Auckland Bangkok Buenos Aires Cape Town Chennai
Dar es Salaam Delhi Hong Kong Istanbul Karachi Kolkata
Kuala Lumpur Madrid Melbourne Mexico City Mumbai Nairobi
São Paulo Shanghai Taipei Tokyo Toronto

Oxford is a registered trade mark of Oxford University Press
in the UK and in certain other countries

Published in the United States
by Oxford University Press Inc., New York

First published February 1955
Second impression August 1955
Fiftieth anniversary edition with new introduction and
afterword by Chris Stringer 2003

British Library Cataloguing in Publication Data
Data available

Library of Congress Cataloguing in Publication Data
Data available

ISBN 0–19–860780–6

1

Typeset in Adobe Garamond
by RefineCatch Limited, Bungay, Suffolk
Printed in Great Britain by
Clays Ltd, St Ives plc

Contents

¤   ¤   ¤

Introduction to the fiftieth anniversary edition of *The Piltdown Forgery*

¤ ¤ ¤

On 21 November 1953, one of the most fascinating puzzles in science was finally solved. This was the day when an article on 'Piltdown Man' appeared in an issue of the *Bulletin of the British Museum (Natural History)*. To great accompanying media interest, and even questions in the Houses of Parliament, the authors—Joseph Weiner, Kenneth Oakley, and Wilfrid Le Gros Clark—described their investigations into the important fossilized human remains found at Piltdown in Sussex in the early 1900s. Their conclusion was stunning: the Piltdown jawbone, and the accompanying materials that supposedly verified this as an ancient fossil, had been faked!

The notorious Piltdown affair probably had its roots in the discoveries of 'Java Man' in 1891 and 'Heidelberg Man' in 1907. These specimens were claimed as missing links between apes and humans, and they may have planted the idea of creating an even more spectacular find on British soil. Charles Dawson, a solicitor and amateur fossil hunter, claimed that some time before 1910 a workman handed him a dark-stained and thick piece of human skull. This had come from gravels at the village of Piltdown. By 1911, Dawson had collected more of the skull from around the site, and had contacted his friend Arthur Smith Woodward, Keeper of Geology at what is now The Natural History Museum in London.

In 1912, Dawson and Woodward began proper excavations at Piltdown, and soon found more skull fragments, some fossil animal bones, primitive stone tools, and a remarkable fragment of a lower jaw. Amid great excitement, they announced the finds to a packed session of the Geological Society in London at the end of 1912, and named a new type of early human, '*Eoanthropus dawsoni*' ('Dawson's Dawn Man'). Although the skull and jaw pieces were awkwardly broken, Woodward reconstructed them into a complete skull that combined a modern-looking braincase with very ape-like jaws. On the basis of the associated animal bones and artefacts, Woodward and Dawson argued that '*Eoanthropus*' was more ancient than Heidelberg Man—equivalent in modern terms to an age of a million years.

Not everyone welcomed Piltdown Man. Some experts, particularly in the United States, were sceptical of the match between the skull and lower jaw, and suggested that they represented separate human and ape fossils that had become mixed in the gravels. In 1913, however, there were more finds at Piltdown, including a canine tooth—intermediate in size between that of apes and humans—and a unique carved artefact made from a large piece of elephant bone. The last Piltdown finds were made in 1915: Dawson supposedly found a molar tooth and some skull pieces closely matching the first finds in a field two miles from the original site, but he never told Woodward the exact circumstances before he died in 1916.

'*Eoanthropus*' became generally accepted as a primitive human fossil, especially in Britain, because it matched

expectations that the brain had evolved to a large size early in
human evolution while other features (such as the jaws and
teeth) had lagged behind. But as early members of the human
family were discovered in Africa and Asia during the 1920s and
1930s, Piltdown Man was pushed into an increasingly per-
ipheral position in the story of human evolution, because
nothing else resembled it.

New chemical and physical dating techniques were
developed after 1945, and these began to be applied to the
fossil record, including Piltdown Man. The first results suggested that the skull and jaw material, unlike the fossil animal
bones from the site, was not very ancient, which made it seem
even more puzzling. Then in 1953, their suspicions aroused,
Joe Weiner, Wilfrid Le Gros Clark, and Kenneth Oakley began
to apply even more stringent tests to Piltdown Man, finally
exposing it as a fake: the Piltdown site had been salted with
bones and artefacts from various sources, most of them artificially stained to match the colour of the local gravels. The
'missing link' itself consisted of parts of an unusually thick, but
quite recent, human skull, and the jaw of an unusually small
orang-utan with filed teeth.

So who was responsible for this hoax, which fooled many
eminent scientists for 40 years? More than 20 different men
have been accused of being involved in the forgery, ranging
from Charles Dawson and Arthur Smith Woodward through to
the eminent anatomists Sir Arthur Keith and Sir Grafton Elliot
Smith. Even Sir Arthur Conan Doyle, the creator of Sherlock
Holmes, who lived in Sussex and played golf at Piltdown, has
been added to the growing list of suspects. But the mystery of
who was really the mastermind behind this extraordinary fake
fossil, and the motive for creating it, remains unsolved.

In the vivid account that follows, Joe Weiner, then an anatomist and anthropologist at the University of Oxford, tells the Piltdown story from the personal perspective of one who knew some of the protagonists, and who played a critical role in its exposure. Of course, as I explain in the Afterword (p. 188), many new theories have been advanced since 1955 about the culprit or culprits behind the hoax, but Joe Weiner's book remains the classic source of information about this enduring mystery.

Chris Stringer
Department of Palaeontology
The Natural History Museum, London
March 2003

Some characters in the Piltdown story

¤ ¤ ¤

ABBOTT, WILLIAM J. LEWIS (1863–1933).
Hastings jeweller and palaeontologist; a member of an influential circle of prehistorians and geologists based at Ightham in Kent.

BARLOW, FRANK (1880–1951).
Technical assistant in Arthur Smith Woodward's Department of Geology at the British Museum (Natural History).

BUTTERFIELD, WILLIAM RUSKIN (1872–1935).
Librarian at the Hastings Museum during the period of the Piltdown discoveries.

DAWSON, CHARLES (1864–1916).
Lewes solicitor and amateur antiquarian, discoverer of many of the Piltdown remains.

DOYLE, (SIR) ARTHUR CONAN (1859–1930).
Doctor and novelist, author of *The Lost World* and creator of Sherlock Holmes.

HEWITT, JOHN T. (1868–1954).
Professor of Chemistry at Queen Mary College, London; served with Samuel Woodhead on the Council of the Society of Public Analysts.

HINTON, MARTIN A. C. (1883–1961).
At the time of the Piltdown discoveries, a knowledgeable volunteer in Woodward's department at the British Museum (Natural History), later Keeper of Zoology there.

KEITH, (SIR) ARTHUR (1866–1955).
Conservator of the Hunterian Museum of the Royal College of Surgeons at the time of the Piltdown discoveries; was involved in crucial discussions of their significance.

LE GROS CLARK, (SIR) WILFRID (1895–1971).
British anatomist at the University of Oxford, one of the three authors of the 1953 exposure of Piltdown.

OAKLEY, KENNETH P. (1911–81).
Geologist, palaeontologist, and archaeologist at the British Museum (Natural History) from 1935; his relative dating analyses of Piltdown presaged the exposure of the forgery.

SMITH, (SIR) GRAFTON ELLIOT (1871–1937).
Professor of Anatomy at the University of Manchester during the Piltdown discoveries; was involved in interpreting the endocranial morphology of the remains.

SOLLAS, WILLIAM JOHNSON (1849–1936).

Professor of Geology at Oxford; considered a supporter of Woodward by some, an enemy by others.

TEILHARD DE CHARDIN, PIERRE (1881–1955).

French Jesuit priest and palaeontologist, he was studying for his ordination when he first met Dawson and Woodward, and excavated with them, discovering the Piltdown canine tooth.

WEINER, JOSEPH S. (1915–82).

South African-born anatomist and anthropologist, he was a colleague of Le Gros Clark's at Oxford, to whom he confided his suspicions about Piltdown in July 1953.

WOODHEAD, SAMUEL ALLINSON (1862–1943).

A chemist at Uckfield Agricultural College and an associate of Charles Dawson.

WOODWARD, ARTHUR SMITH (1864–1944).

Palaeontologist and Keeper of Geology at the British Museum (Natural History); he began excavating at Piltdown in 1912, a few months after Charles Dawson informed him of the first discoveries.

Preface

¤ ¤ ¤

At a recent lecture on the Piltdown disclosures a member of the audience remarked, 'When I read in the paper that Piltdown man was bogus, I felt as if something had gone out of my life; I had been brought up on Piltdown man!' In this remark there is reflected one reason for the widespread interest in the Piltdown affair. For many people of these last two generations Piltdown man represented more or less Darwin's 'missing link', and of course he was all the better known for being a fossil from British soil even if in the world of palaeontology the 'ape man' from Sussex was hardly anything more than an uncomfortable British compromise.

The effect of the scientific exposure on 21 November 1953, announced in the *Bulletin of the British Museum (Natural History)*, was naturally to provoke much speculation as to the identity of the clever author of the hoax. We ourselves had carefully refrained in our public utterances from going into this issue. Not surprisingly, these speculations soon involved the names of almost everyone, dead or living, who might in any way have been concerned in the discovery.

In the course of the scientific investigation of the hoax, I had collected a good deal of material on the personal background of the story, and after the public disclosure more information was brought to my notice. It seemed worthwhile, therefore, to put together this information so as to illuminate as far as possible the problem of the authorship and to dispel various unjustifiable allegations.

I have presented the history of the Piltdown affair more or less as it unfolded itself to me in the course of the year in which I was concerned with it. Thus the problem is first presented in its historical setting and this leads to the scientific investigations which were needed to expose the deception. With this background it is possible to go on and tell what is known of the personalities in the Piltdown affair. There is no reason why more information, and vital information, should not yet be forthcoming. Fully conscious of this possibility I have attempted to treat the evidence with circumspection. The final picture, of course, represents only my own personal interpretation.

From Professor W. E. Le Gros Clark, F.R.S., and Sir Gavin de Beer, F.R.S., and members of their scientific staffs in the Department of Anatomy, Oxford, and the British Museum (Natural History) respectively, I have received much assistance, advice and encouragement. Dr. Kenneth Oakley, of the British Museum, whose pioneer researches have cleared up, besides Piltdown Man, other outstanding problems in the fossil record, has aided me in innumerable ways. My debt to him will be clear from the narrative. Much of the important evidence I owe to the kindness and interest of Mr. W. N. Edwards, Keeper of Geology, who with great patience and perspicacity followed up the information which was offered to him after November 1953. Many of my inquiries were greatly furthered by the enthusiasm of my colleagues Mr. Geoffrey Harrison, Lecturer in Physical Anthropology in the University of Liverpool, and Dr. D. F. Roberts, University Demonstrator in Physical Anthropology at Oxford.

For technical and historical information I am indebted to Mr. J. E. Manwaring Baines, Curator of the Public Museum, Hastings, Mr. Clifford Musgrave, Director, Art Gallery and Museum, Brighton, Mr. N. E. S. Norris, Curator, Sussex Archaeological Society, and Mr. L. V. Grinsell, Head of the Department of Archaeology, City Museum, Bristol. Mr. R. L. Downes, of the Faculty of Commerce, Birmingham University, has aided me considerably in connection with Sussex ironwork and kindred matters of which he is making a special study.

For information and the answering of specific inquiries I am indebted greatly to Sir Arthur Keith, F.R.S., Professor Teilhard de Chardin, Lady Smith Woodward, Mr. L. F. Salzman, F.S.A., Captain Guy St. Barbe, Miss Mabel Kenward, Mr. A. P. Pollard, Mr. E. J. and Mrs. Olivia Lake, Mr. M. A. C. Hinton, F.R.S., Mrs. S. A. Woodhead and Dr. L. S. F. Woodhead, M.B.E., Mr. F. H. Edmunds, Mr. F. W. Thomas, the Rev. Canon Sir Percy Maryon-Wilson, Bt., Mr. R. S. Essex, Mr. F. W. Steer, F.S.A., Mrs. Sonia Cole, the Rev. S. G. Brade-Birks, F.S.A., Mr. P. S. Spokes, F.S.A., Mrs. and the late Mr. E. Clarke of Lewes, Major G. Wade, of Farnham, Mr. A. J. Smith of Leamington Spa, Dr. H. Hensel of the University of Heidelberg, Professor L. C. Eiseley of the University of Pennsylvania, Dr. Patrick C. J. Nicholl of Lewes, Mr. Edward Yates, F.S.A., Dr. E. I. White, Professor H. H. Swinnerton; the Editor of the *Sussex Express and County Herald*; Mr. Charles Gerrard; Messrs. Hampton & Sons Ltd., St. James's, London.

For permission to reproduce the Piltdown portrait and other material in their possession I express my gratitude to the Geological Society of London.

I express my great gratitude for the secretarial work to Miss B. Essex-Lewis and Mrs. D. Forty, and for the photographs to

Mr. F. Blackwell of the Department of Human Anatomy, Oxford, and Mr. C. Horton of the British Museum (Natural History). Finally, for many helpful suggestions I am deeply indebted to my wife and to my friends John and Patience Bradford.

J. S. W.

October 1954

Note to second impression

One important correction only, on page 168, appears in this second impression. Further correspondence with Mr. Hinton makes it clear that the late Mr. Kennard, who said he knew the perpetrator, did not divulge the name, nor did he specifically absolve Dawson. Over the purchase of Castle Lodge, I had earlier supposed that Dawson had opened negotiations on behalf of the Sussex Archaeological Society, but Mr. Salzman stated in the *Sunday Times* that Dawson was not at any time acting for the Society, though the vendors appear to have thought so.

I have tried to make clearer (on page 174) that the artificial treatment which the cranium has received makes it impossible to regard it as of any real antiquity and though close dating is out of the question, there is no reason to suppose that it is more than a few thousand years old: the point has been dealt with by Dr. Oakley in the *Archaeological News Letter* for December 1954, page 125.

1955   J. S. W.

Illustrations

¤ ¤ ¤

1. The Dawn Man of Sussex—*Eoanthropus dawsoni*
   (A composite reconstruction of brain-case and jaw.)

2. The gravel pit at Barkham Manor, Piltdown, Fletching, Sussex. The
   lowest layer is the Tunbridge Wells sands and above it are the gravel
   deposits in which the specimens were found.
   (*Reproduced by permission of the Geological Society of London.*)

3. Personalities Concerned with the Piltdown Discovery
   Back Row: Mr. F. O. Barlow, Prof. G. Elliot Smith, Mr. C. Dawson
   and Dr. Arthur Smith Woodward.
   Front Row: Dr. A. S. Underwood, Prof. Arthur Keith, Mr. W. P.
   Pycraft and Sir Ray Lankester.
   (*From the portrait painted by John Cooke, R.A., in* 1915. *Reproduced
   by permission of the Geological Society of London.*)

4. The inner aspect of the Piltdown mandible (*lower*) is shown for
   comparison with a mandible from a female orang-utan (*upper*); the
   upper specimen has been broken and the teeth abraded to simulate
   the Piltdown specimen.
   (*Photo of the Piltdown mandible by C. Horton. By permission of the
   British Museum* (*Natural History*).)

5. The cast of the Piltdown canine (*left*) is compared with a canine
   from a rather more mature orang; the latter tooth has been stained
   and abraded to simulate the Piltdown specimen.

6. Details of the worked end of the Piltdown 'bone implement' on the

right for comparison with a fossil bone from Swanscombe (*left*) whittled with a steel knife by Dr. K. Oakley to show that the Piltdown fossil bone could only have been worked in this way in modern times.
(*Photo: C. Horton. By permission of the British Museum* (*Natural History*).)

7. 'Stegodon' from Piltdown
Above is a section of a fragment of one of the Piltdown elephant teeth; below is the auto-radiograph obtained by contact with a very sensitive film; it reveals the extraordinarily high radio-activity of the dentine and cementum layers—an intensity not present in any of the other Piltdown specimens or in any comparable specimens from Britain or Western Europe.
(*By permission of the British Museum* (*Natural History*) *and the Geological Survey.*)

8. The Excursion of the Geologists' Association to Piltdown, on 12 July 1913.
(*By courtesy of Mr. Edward Yates, F.S.A.*)

9. The Piltdown Flints (*palaeoliths*). The third specimen from the left is E.606, the staining on which contains chromium. The fourth specimen with inscriptions, came from the collection of the late Mr. Harry Morris of Lewes.
(*Photo: C. Horton. By permission of the British Museum* (*Natural History*).)

1

A Darwinian Prediction

¤ ¤ ¤

Men were on earth while climates slowly swung.
Fanning wide zones to heat and cold, and long
Subsidence turned great continents to sea,
And seas dried up, dried up interminably.
Age after age; enormous seas were dried
Amid wastes of land. And the last monsters died.
                    J. C. SQUIRE: *The Birds.*

On 18 December 1912 Arthur Smith Woodward and Charles Dawson announced to a great and expectant scientific audience the epoch-making discovery of a remote ancestral form of man—The Dawn Man of Piltdown. The news had been made public by the *Manchester Guardian* about three weeks before, and the lecture room of the Geological Society at Burlington House was crowded as it has never been before or since. There was great excitement and enthusiasm which is still remembered by those who were there; for, in Piltdown man, here in England, was at last tangible, well-nigh incontrovertible proof of Man's ape-like ancestry; here was evidence, in a form long predicted, of a creature which could be regarded as a veritable confirmation of evolutionary theory.

Twenty years had elapsed since Dubois had found the

fragmentary remains of the Java ape-man, but by now in 1912 its exact evolutionary significance had come to be invested with some uncertainty and the recent attempt to find more material by the expensive and elaborate expedition under Mme. Selenka had proved entirely unsuccessful. Piltdown man provided a far more complete and certain story. The man from Java, whose geological age was unclear, was represented by a skull cap, two teeth, and a disputed femur. Anatomically there was a good deal of the Piltdown skull and, though the face was missing, there was most of one side of the lower jaw. The stratigraphical evidence was quite sufficient to attest the antiquity of the remains; and to support this antiquity there were the animals which had lived in the remote time of Piltdown man; there was even evidence of the tool-making abilities of Piltdown man. In every way Piltdown man provided a fuller picture of the stage of ancestry which man had reached perhaps some 500,000 years ago.

Dawson[1] began by explaining how it came about that he had lighted on the existence of the extremely ancient gravels of the Sussex Ouse:

'I was walking along a farm-road close to Piltdown Common, Fletching (Sussex), when I noticed that the road had been mended with some peculiar brown flints not usual in the district. On inquiry I was astonished to learn that they were dug from a gravel-bed on the farm, and shortly afterwards I visited the place, where two labourers were at work digging the gravel for small repairs to the roads. As this excavation was situated about four miles north of the limit where the occurrence of flints overlying the Wealden strata is recorded, I was much interested, and made a close examination of the bed. I asked the workmen if they had found bones or other fossils there. As they did not appear to have noticed anything of the sort, I urged them to preserve anything that they might find. Upon one of my subsequent visits to the pit, one of the men handed to me a small portion of an unusually thick human parietal bone. I immediately made a search, but could find nothing more, nor had the men noticed anything else. The bed is full of tabular pieces of iron-stone closely resembling this piece of skull in colour and thickness; and, although I made many subsequent searches, I could not hear of any further find nor discover anything—in fact, the bed seemed to be quite unfossiliferous. It was not until some years later, in the autumn of 1911, on a visit to the spot, that I picked up, among the rain-washed spoil-heaps of the gravel-pit, another and larger piece. . . .

As geologist, Dawson described the formation of these gravels, none of which had been mapped or previously recorded, giving a detailed account of the different strata from which the fossil remains of man and fauna and the tools must have come. He dealt with the question of the chronological age of the gravels and whether all the bones were of the same age, concluding that Piltdown man and some of the mammals were of Early Ice Age, while others were probably older. They represented the remains from an earlier time (the Late Pliocene)[2] which had been washed into the gravels. The gravel itself was composed of layers corresponding to these different ages.

As archaeologist, Dawson gave an account of the salient features of the flint implements. Of these there were two sorts, the 'palaeoliths' which were patently of human manufacture, of an early technique reminiscent of the 'Pre-Chellean' style and technically in accordance with the geological date of the human remains. The other flints, much more abundant, were of doubtful manufacture: they belonged to the class of 'eoliths', flints so crude that archaeologists were acutely divided on the question of their human authorship.

Then Arthur Smith Woodward presented the anatomical description of the animal and human material. Nearly all the animals were represented by fragments of teeth, and these Woodward identified, giving his reasons in detail. Contemporaneous with Piltdown man he concluded were hippopotamus, deer, beaver, and horse. More ancient than the Piltdown man were the remains of elephant, mastodon, and rhinoceros. The Piltdown skull came in for a very detailed examination. Woodward dealt with each cranial piece in turn, and explained how they had been fitted together to give the reconstruction of the complete cranium which was there on view (Plate 1). It had been built up from the nine pieces of cranium and the piece of mandible already unearthed. The striking feature of the cranium was its unusual thickness.

The fragment of lower jaw with the first and second molar teeth still in place obtained, as it deserved, the most careful and systematic description. The shape and size, the markings and ridges for the muscle attachments, the curvature and construction of the specimen, all these, feature by feature, came under scrutiny and led Woodward to his main conclusion: 'While the skull is essentially human . . . the mandible appears to be that of an ape, with nothing human except the molar teeth.' Woodward emphasized in particular those features which served to link the jaw and cranium together in a skull of a single individual. The cranium, for all its human resemblances, exhibited a few simian features—and in this he found support from other distinguished anatomists, while the jaw, ape-like though it was, displayed in the wear of the molars 'a marked regular flattening such as has never been observed among apes, though it is occasionally met with in low types of men'. This unique fossil represented by apish jaw and human brain-case, he was satisfied, merited its own place in the zoological scheme. He therefore proposed its allocation to a new genus and species of man, named 'in honour of its discoverer, *Eoan-thropus dawsoni*'.

At this long-remembered meeting of the Geological Society there was acclaim for Dawson for his part in noticing the gravel pit, for recognizing its great antiquity, and for keeping a constant watch for fossils for many years. There were some who thought that the date which he, as the geologist and archaeologist of the team, had assigned erred on the side of modernity. They urged that a still older date as far back as the Pliocene was indicated, but Dawson gave good reasons for his conservative estimate. Of the extreme antiquity of Piltdown man there was no doubt in anyone's mind. The early Ice Age seemed an entirely reasonable date of emergence for this very early ancestral form, a 'paradox of man and ape' as the creature from Piltdown undoubtedly appeared to be. That his brain had advanced more rapidly than his face and jaw was precisely in accord with current ideas.[3] It was all just as many in the audience had expected. Many there had heard and been convinced by the fervent lectures of Thomas Henry Huxley on the ape-like affinities of man, and Darwin himself in *The Descent of Man* had painted a picture of the earliest human ancestor, the males with 'great canine teeth, which served them as formidable weapons'. 'That we should discover such a race, as Piltdown, sooner or later, has been an article of faith in the anthropologist's creed ever since Darwin's time', wrote Keith.[4] 'On the anatomical side', declared another authority,[5] 'the Piltdown skull realized largely the anticipation of students of human evolution.' The palaeontologist Sollas certainly expressed the prevailing view when he wrote:[6] 'in *Eoanthropus dawsoni* we seem to have realized a creature which had already attained to human intelligence but had not yet wholly lost its ancestral jaw and fighting teeth'. It was 'a combination which had indeed long been previously anticipated as an almost necessary stage in the course of human development'. And finally, Elliot Smith[7] declared the brain of *Eoanthropus*, as judged by the endocranial cast, to be the most primitive and most ape-like human brain yet discovered.

Yet there were a few, at that first meeting, who could not agree with Woodward and Dawson. David Waterston, Professor of Anatomy at King's College, one of the six privileged speakers in the general discussion, found it hard to conceive of a functional association between a jaw so similar to that of a chimpanzee and a cranium in all essentials human.[8] He found it difficult to believe that the two specimens came from the same individual. He and a few others took the view that two distinct fossil creatures had been found together in the gravel. Indeed, those who could not believe that the jaw bone belonged to the skull agreed that the jaw, like the 'Pliocene' group of mammalian fossils—mastodon, elephant and rhinoceros—had been washed into the Piltdown gravel from an earlier geological deposit, whereas the braincase belonged to the later group of Pleistocene fossils like beaver and red deer.

But Woodward's case was coherent and convincing. The creature did fulfil evolutionary expectations in his form, in his age, his tools, and in the character of the animals of the time. Woodward pointed out that the remains had been found very close together, how similar they were in colour and apparently in mineralization, how complementary they were to one another, and how they were functionally connected, as testified above all by the inescapable fact that in this jaw the teeth were essentially human. Their flat wear had never been seen in the molars of apes. It was the sort of wear to be expected from a jaw which was articulated on to a human cranium. That two different individuals were present, a fossil man, represented by a cranium without a jaw, and a fossil ape, represented by a jaw without a cranium, within a few feet of each other and so similar in colour and preservation, would be a coincidence, amazing beyond belief.

Arthur Keith, Conservator of the Hunterian Museum of the Royal College of Surgeons, admitted the strength and logic of Smith Woodward's interpretation. In subsequent years he submitted the Piltdown remains to the most searching examination, adjudicating between the two camps which had formed at the very first meeting. His own criticisms at the time concerned mainly the reconstruction of the cranium and to a lesser extent of the jaw, and these reconstructions were to occupy him in protracted controversy for many years.

Keith drew attention to a crucial point: there was no eye-tooth in the jaw, for most of the chin region had been broken away. What sort of canine would such a creature possess? On this point he did not agree with Smith Woodward's opinion. But Smith Woodward was quite definite. If his interpretation was correct, the tooth when found would certainly be somewhat like that of the chimpanzee, but not projecting sensibly above the level of the other teeth, and its mode of wear would also be utterly different from that of an ape. Like the wear on the molars, the canine tooth would be worn down in a way expected from a freely moving jaw such as the Piltdown man must clearly have possessed in view of its association with so human a cranium. The sort of canine he expected could be discerned in the plaster cast which was before the meeting.

It was very clear to those present how much the missing canine would help to decide the issue of the incipient humanity of the jaw.

Throughout that next long season of digging and sieving of 1913, the oft-discussed canine remained the principal object-ive. Little indeed came to light that season, but on Saturday 30 August, at the end of a day which again had so far proved fruitless, the young priest, Teilhard de Chardin, found the canine, 'close to the spot whence the lower jaw itself had been disinterred'.[9] There was jubilation. The Kenwards, tenants of Barkham Manor (Dawson was the Steward) who had followed the fortunes of the search with unfailing enthusiasm, were appraised of the triumph. It was indeed a triumph. The eye-tooth was just what they had hoped for and closely fulfilled Smith Woodward's prediction of its shape, size, and above all of the nature of its wear. As Dawson wrote in 1915,[10] 'the tooth is almost identical in form with that shown in the restored cast'. Dr. Underwood in 1913 also pointed out this remarkable resemblance, in an article in which, for the first time, X-rays of all the teeth were provided. 'The tooth', wrote Dr. Underwood,[11] 'is absolutely as modelled at the British Museum.'

The new facts further strengthened Woodward's position. Piltdown man could now be said with confidence to possess a dentition in a number of different respects human rather than ape-like, and in the X-ray appearance Keith[12] discovered that the roots of the molar tooth were inserted in the bone in the human and not the ape manner.

The next year's excavation at Barkham Manor yielded what Keith called 'the most amazing of all the Piltdown revela-tions'. Digging a few feet from the place where the Piltdown skull had first been found, the workman with Woodward and Dawson exposed a fossil slab of elephant bone which had been artificially shaped to form a club-like implement. It was found in two pieces 'about a foot below the surface, in dark vegetable soil beneath the hedge which bounds the gravel pit'. The clay encrusting the object enabled Woodward to settle its contemporaneity with Piltdown man, to whose kit of stone tools there was added this, the earliest known bone implement.

The finding of the canine convinced many of the sceptics of the rightness of Woodward's interpretation, but not Waterston, whose opinion remained unchanged till his death in 1921. The two camps persisted. Like Waterston, Gerrit Miller,[13] Curator of Mammals at the United States National Museum, preferred to believe that two fossil creatures were really represented in the Piltdown remains and introduced the new name *Pan vetus* for what seemed to him a new fossil form of chimpanzee. His arguments were met by the zoologists of the British Museum,[14] but Miller continued in his disbelief.[15] At this period Woodward's case was very strong and it had the benefit of Keith's powerful advocacy, presented in masterly fashion in the *Antiquity of Man*.

In 1915 the last, and in its way the most conclusive, of the Piltdown discoveries was announced, for Dawson found the remains of yet another individual two miles away.[16] To those who had been prepared to accept the theory (however far-fetched it might appear) that at Barkham Manor somehow two different creatures had become commingled, this new discovery came as a devastating refutation, for it was hard to conceive of so astonishing a coincidence happening yet again. At the second site at Sheffield Park there were, as before, parts of the brain-case and a molar tooth quite like those previously found. From that site came also another tooth of rhinoceros of, at least, lower Pleistocene age and perhaps older.

The news of the second Piltdown man spread rather slowly and was not fully appreciated until the First World War was over. The foremost French anthropologist, Marcellin Boule, changed his views on learning of this new development.[17] Among the Americans, who for the most part had supported the sceptical attitude of Waterston and Miller, there was a process of general conversion to Woodward's belief. A leader of American anthropological opinion, Fairfield Osborn, had stood out against Woodward with great resolution; his change of mind assumed the nature of a religious conversion. He tells in *Man Rises to Parnassus*[18] how he visited the British Museum after World War I in a mood of the greatest thankfulness that the bombs of the Zeppelins had spared the treasure-house of the Natural History Museum and in particular the priceless Piltdown remains. He tells of the hours he spent that Sunday morning with Woodward going over and over the material and all the arguments, and how at last, in the words of the Opening Prayer of his Yale college song, he felt he had to admit: 'Para-doxical as it may appear O Lord, it is nevertheless true.' Direct handling of the material convinced him that he had been too dogmatic in his two-creatures belief. Woodward had, after all, been right, and, like Keith, Osborn was happy to find himself on common ground and reconciled with Arthur Smith Woodward.

There had been a period of coolness, and indeed, hostility, between Keith and Woodward. Keith admitted the fault lay partly in himself and arose from a feeling of resentment that the unique fossils had not come to him,[19] an established human anatomist, a recognized authority on the skeleton of man and apes, and the Conservator of John Hunter's great anatomical museum at the Royal College of Surgeons. Woodward had treated him with coldness, had kept the new discovery secret from him until a bare fortnight before its public announcement, and then had allowed him only a short twenty-minute visit to South Kensington to view the finds from the Piltdown gravel. Keith's differences with Smith Woodward and Elliot Smith were aroused by the (faulty) reconstruction of the brain-case which Woodward exhibited at the Geological Society meeting. This rather painful argument about the cranium probably did something to distract Keith's attention from the problem of the jaw, for he spent much time and ingenuity and made many searching tests in an endeavour to arrive at a really accurate reconstruction of the cranium, so as to get at its real shape and size. To the whole problem of Piltdown man Keith devoted much painstaking and indeed brilliant anatomical analysis, in the course of which he studied with the greatest thoroughness, to the permanent benefit of other workers, all the relics of ancient man available to him. Though intellectually convinced by Woodward's arguments and the evidence, Keith from the first felt some uneasiness. Many times he assessed the strength and weakness of the case and concluded in favour of *E. dawsoni*. But puzzled he remained and his ambivalent attitude to Piltdown man coloured all his pronouncements. In his work he used the plaster casts made by Mr. Barlow of the British Museum, and distributed in April and May of 1913 to the scientific men principally interested—to Elliot Smith, who was working on the brain of Piltdown man as revealed by the cast of the inside of the skull, to Duckworth at Cambridge, and through Teilhard de Chardin to Boule in Paris. Dawson received one and was able to show it to the many inquirers who now flocked to Piltdown and Uckfield, as Mr. Eade, the present chief clerk at the firm of Dawson and Hart, recollects. There it was seen at this time by Captain Guy St. Barbe, a client of the firm, and by another informant.

By 1915 the British anatomists and palaeontologists were generally of one mind and had accepted Woodward's views—though Waterston still stood out. A Royal Academy portrait[20] (Pl. 3) in oils of 1915 shows us the group of men concerned with the evolutionary study of Piltdown man, who now passed into the general histories and encyclopaedias as easily the best-known of the primal ancestors of the human species. In the centre, holding the reconstructed skull, is Keith, as if to symbolize the newly won harmony of view, with Woodward on one side and Elliot Smith on the other. Woodward's assistants, the zoologist Pycraft (he had been concerned in some interesting study of the jaw and refutation of Gerrit Miller) and Barlow, the skilful maker of the casts, are also of the group. The others depicted are Charles Dawson, Ray Lankester, who had been somewhat sceptical over the implements, and Dr. Underwood, who had advised on dental matters.

The season of excavation of 1916 proved completely unsuccessful. There were many helpers, but nothing was found, either human or animal. Dawson had fallen ill towards the end of 1915, and took no part, though Woodward kept in touch with him. His anaemia however led to septicaemia and his condition became steadily worse. He died on 10 August 1916.

In 1917, after correspondence with Mrs. Dawson, Smith Woodward obtained from Dawson's home, before the auctioneers' sale, the fragments known as the Barcombe Mills skull, and these he deposited in the British Museum.

During the next few years Smith Woodward opened up a number of pits in the vicinity of the original excavation. He also watched closely the digging of some foundations near the farmhouse at Barkham Manor. Except for a flint which he took to be a 'pot-boiler' at the latter site and miscellaneous bone fragments of recent animals, nothing came to light. After his retirement Woodward went to live at Hayward's Heath, near Piltdown, in order to search the original site and the fields of Site II at Sheffield Park, but with no success whatever.[21] He occasionally employed one of the local labourers to do a little digging in these excursions. One such expedition, as late as 1931, yielded only a sheep's tooth.

The site of the first excavations was cleared under the auspices of the Nature Conservancy[22] in 1950 and a large new section of the gravel terrace opened up. Everything was carefully sieved and examined,[23] but the many tons of soil and gravel yielded nothing. This re-excavation made possible the exhibition of a demonstration section of the famous strata protected by a glass window. The cleared area was scheduled as a national monument.

---

1. Dawson, C., and Woodward, A. S., 1913, 'On the Discovery of a Palaeolithic Human Skull and Mandible in a Flint-bearing Gravel overlying the Wealden (Hastings Beds) at Piltdown (Fletching), Sussex', *Quart. J. Geol. Soc. Lond.*, **69**, pp. 117–44.

2. Now termed 'Villafranchian' from the name of the formation which geologists recommend should be used to define the earliest stage of the Lower Pleistocene, that is, the beginning of the Period of Ice Ages which began about 600,000 to 1 million years ago (see Leakey, L. S. B., 1953, *Adam's Ancestors*, pp. 16–9, London, Methuen).

3. Elliot Smith, G., 1912, Address to Section H, British Association, Dundee.

4. Keith, A., 1925, *The Antiquity of Man*, 2nd ed., p. 667, London, Williams and Norgate.

5. Duckworth, W. L. H., in Discussion to Dawson and Woodward, 1913, op. cit., p. 149.

6. Sollas, W. J., 1924, *Ancient Hunters*, 3rd ed., London, Macmillan.

7. Elliot Smith, G., Appendix to Dawson and Woodward, 1913, op. cit., p. 147.

8. Waterston, D., in Discussion to Dawson and Woodward, 1913, op. cit., p. 150.

9. Woodward, A. S., 1915, *Guide to the Fossil Remains of Man*, British Museum (Natural History), p. 20.

10. Dawson, C., 1915, 'The Piltdown Skull', *The Hastings and East Sussex Naturalist*, **2**, p. 182.

11. Underwood, A. S., 1913, 'The Piltdown Skull', *Brit. J. Dent. Sci.*, **56**, pp. 650–2.

12. Keith, A., op. cit., p. 684.

13. Miller, G. S., 1915, 'The Jaw of Piltdown Man', *Smiths. Misc. Coll.*, **65**, pp. 1–31.

14. Woodward, A. S., 1917, 'Fourth note on the Piltdown gravel with evidence of a second skull of *Eoanthropus dawsoni*', *Quart. J. Geol. Soc.*, **73**, p. 9.

15. Miller, G. S., 1918, 'The Piltdown Jaw', *Amer. J. Phys. Anthrop.*, **1**, pp. 25–52.

16. Woodward, A. S., op. cit., pp. 1–7.

17. Boule, M., 1923, *Les Hommes Fossiles*, 2nd ed., pp. 158–76, Paris, Masson et Cie.

18. Osborn, H. F., 1927, *Man Rises to Parnassus*, pp. 45–74, London, Oxford Univ. Press.

19. Keith, A., 1950, *An Autobiography*, pp. 324–5, London, Watts.

20. Painted by John Cooke, R.A., and presented to the Geological Society in 1924, by Dr. C. T. Trechmann, F.G.S.

21. Woodward, A. S., 1948, *The Earliest Englishman*, pp. 12–13, London, Watts.

22. Toombs, H. A., 1952, 'A New Section in the Piltdown gravel', *The South-Eastern Naturalist and Antiquary*, **67**, pp. 31–3.

23. By Mr. Toombs, Dr. Oakley, and Mr. Rixon.

2

An Impasse

¤ ¤ ¤

D awson had received widespread recognition, but died too soon to be given any special award from a scientific body. Twenty years later his achievement was commemorated by the erection of a memorial stone at the site of the gravel pit at Barkham Manor. Sir Arthur Smith Woodward had taken the initiative in this and borne most, if not all, of the expense. The unveiling was done, at his request, by his old friend Sir Arthur Keith at the well-attended ceremony on 22 July 1938. Keith gave a brief but eloquent oration. He dwelt on the wonderful achievement of the keen-sighted amateur Dawson, an achievement which he likened in the history of discovery to that of the French lock-keeper, Boucher de Perthes—the first man, three-quarters of a century ago, to recognize clearly the human workmanship of the Ice Age flint hand-axes of the Somme. The discovery at Piltdown ranked worthily, too, with that of Neanderthal man discovered in 1857, the first known of all fossil men. These discoveries had encountered tremendous opposition before acceptance was won. The claims of Perthes had brought incredulity and set the scientific world a momentous problem, and only after years of stormy argument were these claims conceded; the discovery of Neanderthal man likewise brought disagreement and controversy. But this fossil form was accepted in the end. As Keith said, then came Dawson's discovery, and this brought the greatest problem of all. But Keith did not go on to claim that all was now well with 'the earliest known representative of man in Western Europe', of which he had just finished a laborious re-study. A puzzle it had always been and a puzzle it was still.

Keith could not hide his underlying doubt, and ten years later he expressed it again in the Foreword which he wrote at Lady Smith Woodward's request to Woodward's own book, *The Earliest Englishman*, published posthumously in 1948. He declared: 'The Piltdown enigma is still far from a final solution.'

Why should Keith still express such doubt and bewilderment? But it was no longer surprising. By 1948 there were many who had decided that little sense could be made of the Piltdown puzzle. For the many new palaeontological discoveries of the previous decade had left the Dawn Man completely isolated and without any certain affinities in the broad stream of human evolution.

It must be remembered that when Woodward and Dawson made their momentous announcement in 1912 the number of fossils which had any bearing on the early stages of human evolution could be counted on the fingers of one hand. The most primitive of these was Java man. Hailed when it was found in 1891 as *the* missing link, even its discoverer, Dubois, began to have doubts whether he had in fact correctly described the creature as an ancestral human being. He was inclined to think that Java man was only an extinct form of giant gibbon. Nor was the geological date at all clear. Then, as already mentioned, there was the Heidelberg (or Mauer) jaw of 1907, a completely isolated jaw from a time seemingly contemporary with Piltdown.[1] Apart from the Java fragments and the Heidelberg jaw, there were only the much later specimens of the bigger-boned and rather bestial-looking Neanderthal (Mousterian) man. There was thus little enough with which to compare *Eoanthropus dawsoni*, whose claims appeared in the circumstances all the more convincing and cogent. In this dearth of fossils, Woodward could quite reasonably hypothesize that 'surviving man may have arisen directly from the primitive source of which Piltdown skull provided the first discovered evidence'. 'Piltdown man, or some close relative' is 'on the direct line of descent with ourselves'. A not unreasonable picture, in the circumstances of the time.

But the picture changed as discoveries accumulated from China, Java, and Africa. About 1936 a whole series of new finds of Java man were made. One of these was almost a duplicate of Dubois' first specimen, and thoroughly vindicated the original claims. *Pithecanthropus* was an undeniable primitive hominid; his skeletal features were settled beyond any doubt. And by that time a closely allied creature from Peking, probably a slightly more advanced hominid, was known from the remains of a score or more individuals—cave-dwelling, firemaking, and tool-using primitive men. By 1948, when Keith wrote his worried Foreword to Woodward's posthumous book, a still earlier pre-human stage in man's ancestry had been recognized by Dart and Broom in the caves of the Transvaal. These were the Australopithecinae, very ape-looking, but with many marks of an incipient humanity.

From all this there emerged a picture of human evolution quite different from that which had been worked out from the interpretation of the Piltdown material. Contrary to the beliefs of the Piltdown supporters, all these other fossils agreed in showing Man as having obtained his large brain only slowly, whereas many features of jaws and teeth became human very early on. These early forms had a chin region, teeth, and a general shape of jaw which like that of modern man differed basically from those of the ape or Piltdown. Where Piltdown had an extremely modern forehead and an ape's jaw, Java and Peking man possessed the combination in reverse—a simian looking forehead and an unapelike jaw.

It had now to be concluded, and Woodward himself did so in 1944, that two quite separate evolutionary lines existed. Along one went the South African, Java, Peking, Neanderthal sequence—an overlapping series of transformations; alone on the other was Piltdown man. The two lines were irreconcilable. No common ancestor for the two lines was in sight, and now much argument was heard as to which line had given rise to *Homo sapiens*. Many, including Woodward and Keith, still favoured the Dawn Man.

This was a rather complex state of affairs, but by no means an impossible one—though, of course, it was made still more complicated by those who still believed in the existence of two fossil Piltdown creatures, fossil man and fossil ape; but they, however, could add no conclusive evidence on this point nor demolish Woodward's case.

Complete confusion succeeded when the geologists decided that the early date of *Eoanthropus* could not possibly be correct. The result of the fluorine dating test announced by Dr. Kenneth Oaklay[2] in 1949 brought about this decisive change of outlook.

In 1892 a French mineralogist, Carnot, reported that the amount of fluorine in fossil bones increases with their geological age; the fluorine present in the soil water steadily accumulates in the bones and teeth. When Dr. Oakley some years ago came across this entirely neglected paper he realized[3] that the method would serve to establish whether bones found close together in a single deposit were of the same or of different ages. It seemed just the method to apply to the Piltdown bones to see whether the human remains were as old as the palaeontologists still claimed; at the same time (and even more important) the fluorine test should show whether the jaw and cranium were really of the same age. The method had worked successfully in other disputed cases.[4] Now used on the Piltdown remains the test produced its greatest surprise.

The fluorine content of jaw and cranium turned out to be remarkably low (0·1–0·4 per cent.), far less than that of the Early Ice Age (and even earlier) mammals of the Piltdown deposit, and similar to the deer or the admittedly late beaver. The fluorine values ranged from less than 0·1 to 0·4 per cent. for the various fragments of the two Piltdown men as compared with about 2 per cent. for *Elephas* and *Mastodon*. There was here no question of a vast antiquity for Piltdown man; in fact, the fluorine accumulation was so small that the bones could be given a dating hardly later than the Upper Pleistocene, i.e. the last part of the Ice Age, 'probably at most 50,000 years old'. Nor did this radical change of date through the fluorine test go unsupported, for attention had already been drawn by Dr. Oakley[5] to F. H. Edmunds' revision of the Geological survey map (of 1926) of the Ouse near Uckfield, which indicated that the actual gravel terrace[6] had all along been mistakenly attributed to a higher and much older level.

This was startling. It was startling both to those who believed that the remains were those of a single creature (the monistic view) as well as to those of the opposing (or dualistic) view who held that two fossil creatures were present.

A Dawn Man as late as this new date was an anomaly indeed. His title to human ancestry lapsed at once; not only were there at this revised date many examples of fully developed men of our own type in existence,[7] some representatives had even made their appearance long before.[8] On the old date Piltdown man would have left a long line of descendants, and some of these in the time available might conceivably have undergone considerable changes, even perhaps attaining to the form of *Homo sapiens*, as Woodward thought. But on this late dating what descendants could he possibly have? And as for ancestors, this late surviving monstrous creature could not be linked at all with the earlier men of Java or China. On this new view Piltdown man had neither visible ancestors nor descendants. The advocates of this 'monistic' view could only suppose that the Dawn Man was after all an extreme specialization, a divergent line which had led to nowhere—which was saying in effect that his evolutionary importance was modest indeed.

But what of the 'dualists'? To them the revised date of the remains brought its own complexities. For the dualists the cranium now merely ranked as another of the many fossil specimens of *Homo sapiens* of the Late Ice Age—no different essentially, for example, from the cave-dwelling men of Magdalenian culture who were responsible for the magnificent cave art of Southern France and Spain. It apparently escaped them that this produced the anomaly of attributing the poorly worked flints to a much more advanced individual. As for the jaw, which the 'dualists' were inclined to regard as older than the brain-case, it would now have to be attributed to a fossil ape living in England near the end of the Ice Age in company with the beaver—a wholly improbable event in view of the climatic vicissitudes and the absence of great apes throughout the Ice Age in Europe anywhere. And what of all Woodward's evidence linking the jaw with the cranium? That had not been eliminated. But what made matters worse for the dualists was the fact that the fluorine content of jaw and cranium were quite similar, a finding which if anything increased the probability that the skull fragments all belonged to the same individual. As Professor Straus has written,[9] 'the abolition of a Lower Pleistocene dating did not solve the Piltdown problem. It merely produced a new problem that was even more disturbing.' The 'new' Piltdown problem of 1949 was more than that. It had become a mystery, bewildering to 'dualists' and 'monists' alike.

The impasse revealed itself very soon in the writings of this period, and a chaotic set of opinions began to enliven the textbooks. The 'dualists' gained perhaps most support, but only in disregard of all the anatomical and other evidence, of the fluorine results, and of the existence of the second Piltdown man. The 'monists' accepted the Dawn Man in his isolated condition, perched far at the end of his evolutionary branch; there was even a far-fetched suggestion that the jaw was not really ape-like, if properly reconstructed! There were those who advocated 'neutrality' and hoped for new evidence one day to decide between the two issues (though neither offered a real solution!). Some suggested that the Piltdown discovery should be entirely disregarded and the man of Piltdown cast out of the evolutionary calendar.

---

1. Dawson, C., and Woodward, A. S., 1913, op. cit., p. 137.
2. Oakley, K. P., and Hoskins, C.R., 1949, *Abstr. Proc. Geol. Soc. Lond.*, 14

December 1949; full details: 1950, 'New Evidence on the Antiquity of Piltdown Man, *Nature*, **165**, pp. 379–82.

3. Oakley, K. P., 1948, 'Fluorine and the Relative Dating of Bones', *Advancement of Science*, **16**, pp. 336–7.

4. Oakley, K. P., and Montagu, M. F. A., 1949, 'A Reconsideration of the Galley Hill Skeleton', *Bull. Brit. Mus. (Nat. Hist.) Geol. Sect.*, **1**, pp. 27–46.

5. Oakley, K. P., 1937, *J. Roy. Anthrop. Inst.*, **67**, p. 394.

6. Edmunds, F. H., 1926, in *Geology of Country Around Lewes*, Mem. Geol. Surv., 1926, pp. 63–8.

7. Examples are the fossils from Skhūl (Palestine) and Fontéchevade.

8. Swanscombe man, for example.

9. Straus, W. L., Jr., 1954, 'The Great Piltdown Hoax', *Science*, **119**, pp. 265–9.

3

An Hypothesis

¤ ¤ ¤

Towards the end of July 1953 a congress of palaeontologists was held in London under the auspices of the Wenner-Gren Foundation. The problems of fossil man were the subject of its deliberations. Java man, Neanderthal man, Rhodesian man, the South African prehumans—all these were given close attention. But Piltdown man was not discussed. Not surprisingly. He had lost his place in polite society. What more could one usefully say about him? Yet, unofficially, the Dawn Man did manage an appearance. Most of those present had not seen the original fossil specimens, so on a tour of the Natural History Museum these were shown along with others housed there. The sight of the actual fragments provoked the familiar tail-chasing discussion. As always there were those who could not feel that the famous jaw really harmonized with the rest, but there were others who took the opposite view. The enigma remained.

At the dinner that night Dr. Oakley remarked casually to Dr. Washburn of Chicago and myself that owing to Dawson's early death in 1916 the Museum had no record of the exact spot where the remains of the second Piltdown had been found. They knew the place—Sheffield Park—but the actual spot or even the field had never been marked on a map. 'The fact is', said Oakley, 'that all we know about site II is on a postcard sent in July 1915 by Dawson to Woodward, and an earlier letter in that year, from neither of which can one identify the position of Piltdown II.' This was surprising. The second group of finds had done so much to convince many people that the first Piltdown man was by no means an isolated phenomenon. One had imagined that if it were ever thought worthwhile it would be possible to go and excavate the second site. Now it appeared that this had never been done because the second site could not be located, though Woodward had apparently visited it before the second find. This curious piece of information greatly puzzled me. I knew that Dawson had died in 1916 and it seemed difficult to understand why he had not recorded so important a fact as the location of Piltdown II. Dawson had a reputation, I knew, for great conscientiousness and accuracy. Sir Arthur Keith had spoken in the highest terms of his qualities. Perhaps Woodward had been told verbally and somehow his own record had been mislaid? I did not know then that Dawson had been ill for nearly a year before his death.

This small puzzle turned my thoughts to the larger Piltdown conundrum. My own conclusion when reviewing the matter in 1950,[1] like that of others, was that Woodward's *Eoanthropus* had become a complete anomaly, that the only course was to wait till more material was dug up, and that it was really profitless to spend much time on choosing between possibilities, none of which was susceptible of final proof. Thinking it all over again, I realized with astonishment that while there were in fact only the two possible 'natural' theories, i.e. that Piltdown man was in fact the composite man-ape of Woodward's interpretation, or that two distinct creatures, fossil man and fossil ape, had been found side by side, neither of the 'natural' explanations was at all satisfactory. If the two 'natural' explanations failed in some way or another, what other possible explanation could there be? Was there any other way of resolving the whole disorder of fragments, dates, chronology? On evolutionary grounds alone a late Dawn Man stood out as an obvious incongruity. The riddle might be approached more simply (I argued) by accepting at once the extraordinary difficulties of regarding the fossil as an organically single individual and by concentrating entirely on the perplexities of the two-creature hypothesis. What were feasible alternative explanations of the coincidence of two distinct individuals? If the jaw and cranium had not come together by nature or by blind accident then could they have got there by human agency? This would mean that someone by mischance or error had dropped a fossil jaw in the pit (perhaps used as a rubbish dump) dug in gravel which happened to contain other fossil remnants.

But surely this could hardly have been repeated at the second site? Perhaps site II was after all of exaggerated significance or had been mistakenly interpreted. As Hrdlička[2] and others had been saying all along, perhaps the single molar might not really be ape or have any affinity whatever with the first teeth, so that the Sheffield Park fragments, despite their other similarities to those at Piltdown, would simply be a quite ordinary set of human remains. Or we could dispose of Piltdown II by supposing that the bits had actually come from the Barkham Manor site two miles away, in gravel brought across for some reason or other. Even if one were prepared to accept them, this elaboration of ancillary hypotheses still avoided the main issue. For even if the jaw had been thrown on the gravel, to meet with the cranium, it was still a *fossil* jaw and we had not in fact escaped the original dilemma: what fossil ape could it possibly be? Still, the idea of an accidental deposit or loss of a jaw could be pursued a stage further (still disregarding site II) if we postulated that the jaw was not a fossil, but really that of a *modern* ape. We might then accept the accidental coincidence. But could the jaw possibly be modern? Immediately strong objections loomed up. To say the jaw was modern implied that the fluorine analysis had been inaccurate or that the published results must be in some way compatible with modern bone recently buried. In effect this would imply that the most reasonable interpretation of the results had been in error. That difficulty was dwarfed at once by a far more serious objection. The teeth were almost unanimously acknowledged to possess features quite unprecedented in modern apes—the flat wear of the molars and the curious type of wear of the canine had never been matched in an ape's mandible.

A modern jaw with flat worn molars and uniquely worn-down eye tooth? That would mean only one thing: deliberately ground-down teeth. Immediately this summoned up a devastating corollary—the equally deliberate placing of the jaw in the pit. Even as a mere hypothesis this inference could at once dispose of two of the most intransigent Piltdown posers: how the jaw and teeth had ever got there and how the teeth had come by their remarkable wear. But the hypothesis of a deliberate 'salting' of the Piltdown gravels clearly carried much wider implications, and the idea was repellent indeed. Could one not find a fatal flaw at once, and quickly dismiss this as a solution of the Piltdown mystery? There would be no need to consider the idea any further or even to examine the specimens (or rather the casts) in the laboratory next day. (For this cogitation had occupied the small hours on my return to Oxford after the Wenner-Gren dinner.)

What then were the immediate points of weakness and strength of this theory? After further reflection, the only serious surviving objection seemed to be the figure for the fluorine analysis. It had to be admitted that a modern specimen was rather unlikely to exhibit a fluorine content quite so high as the published figures, even supposing that the ape had lived in a fluorine rich area. Yet even this objection could be countered. Oakley pointed out[3] that the probable error of the method of estimation used in 1949 was actually stated to be possibly as much as ±0·2 per cent.[4] Thus although recorded as 0·3 per cent. the fluorine content of the jaw might in fact be less than 0·1 per cent., as in recent bone. If the fluorine value was not a fatal objection, an *a priori* case for a deliberate hoax assumed some strength on a number of counts. Clearly, if the intention was to pass off as a fossil a modern jaw, it would still not have passed scrutiny, despite the abraded teeth, unless certain tell-tale features were first removed. And it was just such features that were missing. Nearly the whole chin region was lost and only the two molar teeth were left in place, and, more telling, the bony knob where the jaw articulates with the skull had been broken away. This knob (or condyle) would certainly have made it apparent that the jaw would not fit into the cranium. For there is often a marked difference between the human and ape-like articular condyle. Then there were the strange characters of the canine. Deliberate tampering with the tooth would easily explain this particular oddity.

This *a priori* case obtained added support from discussions with Professor Le Gros Clark and our examinations of the Piltdown casts in the Department of Anatomy at Oxford.

Perhaps the most telling argument which could be marshalled at this stage lay in an extraordinary fact revealed by the anatomical reconsideration of the remains. It appeared to me that, despite the many claims advanced from time to time for the existence of a whole variety of human features in the jaw and of ape-like features in the cranium, the only completely acceptable and undoubted characteristic of a human kind in the jaw turned out to be the flat wear. Nothing else could unequivocally be said to be human. How strange, then, that this one feature should be present to link jaw and cranium and yet these were supposed to form a harmonious combination in a live animal. Surely a few other modifications should have been apparent in the jaw. Yet, as Woodward himself had often pointed out, such functional features of the jaw as its muscle attachments were entirely ape-like. The moment one attributed this flat wear to a deliberate abrasion of the teeth it became understandable.

It now appeared from our discussion that the canine was not merely peculiar in its mode of dental wear, but that it was itself paradoxical in that the wear was so heavy as to be quite out of keeping with the immaturity of the tooth. This was a fact first pointed out in 1916 by the dentist Mr. Lyne,[5] and never properly explained. Lyne's cogent arguments had been brushed aside by Woodward and Underwood.

Then we examined the plaster casts. These revealed features quite understandable as the outcome of artificial abrasion of the dental crowns. In particular we were struck by the extraordinary flatness of the second molar and the lack of a smooth continuity of biting surface from the one molar to the next. Next, a chimpanzee's molar, of about the same size as the Piltdown, was experimentally filed down. This proved easy enough to do and, even without any polishing of the surface, by staining with permanganate an appearance very like the Piltdown molars was obtained, as far as could be judged from the casts and from photographs.

Yet another piece of positive information emerged when one re-read Dr. Oakley's fluorine paper.[6] During the course of drilling to obtain his sample of dentine Dr. Oakley had made an observation which now assumed a special significance: 'Below the extremely thin ferruginous surface stain', he had written, 'the dentine was pure white, apparently no more altered than the dentine of recent teeth from the soil.' Re-reading of Dawson's and Woodward's papers further made it clear that they themselves had missed a chance of making what might have been a decisive comparison of jaw and cranium. It appeared that only on the cranium had chemical tests for organic matter and other constituents been made, and the cranium had been found to contain no organic matter. If the jaw was modern, its organic content would be high, but the analysis had not been done.[7]

Though all these points built up quite a strong *prima facie* case, new objections also appeared, to add to that provided by the rather anomalously high fluorine content already mentioned. There were two serious points of criticism. The dental wear of the canine had been pronounced to be indubitably natural by Dr. Underwood at that 1916 meeting, when he spoke in violent disagreement with Mr. Lyne's contention of the immaturity of the canine and its paradoxical nature. Dr. Underwood had pointed out that the X-ray showed clearly a patch of *secondary dentine* such as always is deposited progressively with natural wear. The other difficulty arose from the radiographs of the molars. The relatively short roots were not really ape-like, as Keith had pointed out. They furnished another near-human attribute on the jaw.

Our general hypothesis seemed sufficiently sound, however, to warrant an approach to the authorities at the British Museum for renewed investigations, anatomical, radiological, and chemical, of the Piltdown material. These investigations would be needed for three reasons if the hypothesis was to lead to proof: firstly, for confirmation (or otherwise) of the evidence already gleaned; secondly, to establish the validity (or otherwise) of the various objections to the hypothesis; thirdly, to apply any new tests which might be suggested at this stage.

To these lines of investigations others were added after the publication of the first report[8] and the results of these will be given their place in the narrative. But in August and September of 1953 we already had many critical tests to do. The fact that most of these were of quite independent characteristics, chemical, physical and biological, and on different specimens, meant that agreement between them would amount to overwhelming proof. Equally clear and in view of the serious nature of the 'fraud' hypothesis, there would be every need for a complete and all round agreement in the tests.

The tests we had in mind at this stage turned simply on the issue of the modernity or otherwise of the jaw and teeth, but it was obvious that the implications extended to every aspect of the Piltdown discoveries.

At this time awaiting the outcome of our 'predictions' and repeatedly arguing and reviewing our case, and seeing no other possible solution to the problem, we could well appreciate Holmes's sage advice to Watson:

'How often have I said to you that when you have

eliminated the impossible, whatever remains, *however improbable*, must be the truth.'[9]

---

1. 'Physical Anthropology since 1935', in *A Hundred Years of Anthropology*, by T. K. Penniman, 1952, London, Duckworth.

2. Hrdlička, A., 1922, 'The Piltdown Jaw', *Amer. J. Phys. Anthrop.*, **5**, pp. 337–47.

3. Letter to Professor Le Gros Clark, 12 August 1953.

4. Oakley, K. P., and Hoskins, C. R., 1950, loc. cit., p. 380.

5. Lyne, C. W., 1916, 'The significance of the radiographs of the Piltdown teeth', *Proc. Roy. Soc. Med.*, **9**, pp. 33–62.

6. Oakley and Hoskins, op. cit., p. 379.

7. The Moulin Quignon jaw was declared to be a deliberate intrusion on the evidence of the high nitrogen content, by Falconer and Busk in 1863. See Keith, A., 1925, *Antiquity of Man*, pp. 270–1.

8. Weiner, J. S., Oakley, K. P., and Le Gros Clark, W. E., 1953, 'The Solution of the Piltdown Problem', *Bull. Brit. Mus. (Nat. Hist.) Geol.*, **2**, No. 3, pp. 139–46.

9. Conan Doyle, *The Sign of Four*.

4

The Jaw Displaced

✠ ✠ ✠

On the basis of our preliminary arguments and our anatomical re-examination of the fragments, Mr. W. N. Edwards, the Keeper of Geology of the Natural History Museum, felt justified in allowing the specimens of mandible, cranium, and teeth to be drilled for much larger samples than could ever have been sanctioned hitherto. These larger samples and the use of improved chemical methods guaranteed a high degree of analytical reliability.

The drilling itself gave us an encouraging start. As the drilling proceeded, Dr. Oakley and his assistant perceived a distinct smell of 'burning horn' when the jaw was sampled, but they noticed nothing of the sort with any of the cranial borings. This subjective indication of some distinct difference between the constitution of jaw and cranium soon gained objective confirmation. The drilled sample from the jaw proved to be utterly unlike those from the cranium. In keeping with the belief in its fossil or semi-fossilized character, the latter produced a fine particulate granular powder, whereas the jaw yielded little shavings of bone, just as did a fresh bone sampled as a control. Here was the beginning of the series of findings which progressively widened the gulf between jaw and cranium.

Very soon Dr. Oakley obtained clear chemical evidence to justify fully the strong suspicion of the modernity of the jaw and of the totally distinct origin of the cranium. An improved technique for estimating small quantities of fluorine produced this decisive result. The cranial fragments of site I were found to contain fluorine in a concentration of 0·1 per cent., a value somewhat similar to that of specimens of known Late Ice Age. The jaw and the three teeth on the contrary gave much lower figures, at levels below 0·03 per cent., values well within the range of known modern and fresh specimens. Indeed, these values are on the borderline of the sensitivity of the method. The fluorine test gave its verdict twice over. For the two cranial fragments from the second Piltdown site contained a fluorine concentration of 0·1 per cent. and the isolated molar which went with these fragments contained less than 0·01 per cent. These fluorine results alone go far to settle the main issue. As the reader will recall, the method serves essentially to compare the dates of material from any one deposit, and the Piltdown fluorine values prove not only that the jaw and teeth do not belong to the crania but that they are of younger date, and the test shows this to be true at both Barkham Manor and Sheffield Park.

With this Dr. Oakley and his associates[1] now launched a whole battery of chemical and physical tests at the fragments, bringing to bear on the Piltdown problem an array of new techniques in the last few months of 1953 exceeding all endeavours of this kind in the whole history of palaeontology. In succession they tested and compared the fragments for iron, nitrogen, collagen, organic carbon, organic water, radio-activity and crystal structure. This list is an epitome of the resources which the chemist and physicist have in recent years put in the hands of the archaeologist and palaeontologist, and in the Piltdown problem these methods obtained a thorough trial.

The test for nitrogen content, greatly improved by Cook and Heizer,[2] represented an independent approach for comparing the respective ages of the different fragments. Whereas the fluorine assay reflects the accumulation in bone of an extraneous element, the nitrogen content indicates the progressive loss of organic matter from the bone itself. Thus in fresh or recently buried bones and teeth, the fluorine content is low while the nitrogen is high with values of the order of 4 per cent. With the passage of time, as fluorine accumulates, nitrogen would tend to decrease.[3] The nitrogen results of the Piltdown specimens were quite clear: Piltdown jaw, 3·9 per cent.; Piltdown canine (dentine), 5·1 per cent.; molar tooth (dentine), 4·3 per cent. at site I, 4·2 per cent. at site II; fresh bone, 4·1 per cent.; modern chimpanzee molar, 3·2 per cent.; cranial fragments at site I, 1·4 per cent.; frontal bone at site II, 1·1 per cent.; occipital bone at site II, 0·6 per cent. The findings need little explanation. With the fluorine results, and independently, they prove that at both localities recent or modern jaw and teeth are in association with cranial bones of a different and much older constitution.

Here Dr. Oakley posed a serious objection. Could not these nitrogen values be vitiated by the possibility that in the making of plaster casts gelatine moulds might have been used or that the specimens had been sized? He answered this by pointing out that if such were the case the far more porous cranium would have absorbed as much and probably far more nitrogen than the denser dentine of the teeth, whereas the reverse is in fact what the analyses reveal. The point was settled by arranging with Professor Randall's Unit[4] for examinations (by means of the electron microscope) of the organic, nitrogenous fibrous material (collagen) itself. In keeping with the nitrogen values, an abundance of collagen with the characteristic banded appearance was revealed in the jaw and tooth, and an absence of the material from the cranium. Once again we have a separation into recent and older groups.

Another consideration had to be borne in mind in our investigations. Woodward (1948) said of the mandible: 'It had evidently been missed by the workmen because the little patch of gravel in which it occurred was covered with water at the time of the year when they reached it.' This raised the possibility that conditions in the Piltdown gravel were exceptional and perhaps, through being water-logged, reducing conditions prevailed in the basal bed and had led to the preservation of the collagen. However, investigation[5] disproved this possibility, for the chemical state of the deposits in February 1951 was such that in fact oxidizing conditions were present.

Later on Dr. Oakley obtained estimations of the organic carbon and chemically-bound water and these, like the nitrogen, mirror the organic content. Like the nitrogen, they yield high values in the jaw and teeth, low values in the several cranial fragments.

The fluorine results confirmed by the nitrogen values, as has been shown, suffice to testify to the modernity of the jaw. But a modern ape's jaw would not be the brownish, slightly yellowish-tinged, colour of the Piltdown jaw unless it had been stained either by using chemicals or perhaps by being left for a sufficient time in iron-containing soil or mud such as the Ouse gravel in fact contains. The coloration of the skull fragments and the blackish brown coating of the canine tooth were investigated by a variety of means.

The tests made for iron early in the investigation yielded their own confirmation of what the fluorine and organic analyses had demonstrated. A distinct contrast was apparent in the iron-staining of jaw and cranium. The drill samples showed that the brown colour had penetrated uniformly and evenly through the porous and drier semi-fossilized pieces of skull-cap; in the jaw the staining was quite superficial and a few millimetres below the surface the bone became progressively lighter. There was 7 per cent. iron in the cranium. In the jaw it fell rapidly from about 8 per cent. to 3 per cent. below the surface. All this was, of course, consistent both with the separate identities of the jaw and cranium and the belief that the former represented only a recent burial in the Piltdown gravel. The result raised the suspicion that the staining might have been intentional.

Of deliberate staining we obtained a striking proof when the canine came to be examined. The tooth has a darkish brown outer coat, always taken to be an ordinary iron-stain, and it was under this 'ferruginous' layer that Dr. Oakley, it will be recalled, remarked with surprise the whiteness and freshness of the dentine. But there proved to be only minimal quantities of iron (oxide) in this stain, the nature of which eluded identification for some time. The layer was found to be a paint-like substance forming a flexible film. The possibility that it was a dried-out layer of Chatterton's compound[6] was ruled out, amongst other things, by its low solubility in organic solvents, but like this compound it contained bitumen. Finally, it turned out to be a bitumen earth containing iron oxide, in all probability the well-known paint—Vandyke brown. It might have been argued that bituminous earth could produce a natural incrustation were it not known that bituminous matter is entirely out of place in a highly oxidized gravel. Its artificiality is established beyond doubt by the finding by Dr. Claringbull of a minute spherule of an iron alloy embedded in the coating on the outer (labial) surface of the crown. The reddish brown stain on the occlusal or chewing surface (like that on the molars) is probably also a ferruginous earth pigment applied as an oil paint (e.g. red sienna).

Here we may mention briefly the two rather novel and up-to-date physical techniques which in the first instance were pressed into service in the examination of the skull. They were destined to lead us to findings as astonishing as the demonstration of the falsity of the jaw, and to clarify for us the significance of the colour of the specimens.

When the bones of *Eoanthropus* were tested with a Geiger counter the radio-activity was found to be almost indetectable in the jaw, which is a further confirmation of its modernity. The cranial fragments were slightly but appreciably radio-active, and this is attributable perhaps to different origin but more likely to the use of an oxidizing agent like potassium bichromate.

The other technique involved the X-ray examination of the crystal structure of bone. This crystallographic technique provides a clear identification of the mineral complex of calcium phosphate called apatite of which bone is composed. The bone is powdered and when the X-ray beam is played on the powder there is a characteristic absorption of some of the X-rays and reflection of the others. The resulting X-ray picture is quite specific. The apatite is clearly revealed in the jaw. But in the cranium it is aberrant. The crystallographic picture is that of a complex containing sulphate, and which is allied to gypsum. In other words the jaw contains no sulphate while the cranium does.

We may now put the results[7] together in this simple table:

	Flourine	Nitrogen	Carbon	Water	Collagen	Radioactivity	Sulphate	Iron-stain
Piltdown *jaw* (*and teeth*):	Negligible	Very high	High	High	Present	Negligible	Absent	Superficial
Piltdown *cranium*:	Present	Low	Low	Low	Absent	Slight	Present	Even

The divorce between mandible and skull cap is complete.

The chemical and physical tests had yielded overwhelming proof[8] of the primary contention that the jaw was modern and gave some clear evidence of fraudulent activity. Those tests had disposed of a major objection (the original fluorine values), confirmed the postulated expectations (in the new fluorine values, the nitrogen results, and the iron-staining) and added a whole new series of independent confirmatory evidence (the tests for organic carbon, bound water, radioactivity and the 'apatite-crystal'). The chemistry and physics had done all this virtually twice over, and even more, for the results were consistent at the two sites and on the different fragments at each site.

None of these findings was yet available on 5 August (1953) when Professor Le Gros Clark and I met Dr. Oakley at the British Museum to carry out the anatomical re-examination. The casts at Oxford, as already remarked, had provided some definite indications to favour the belief that the extraordinary occlusal wear of molars and canine represented nothing else but the results of deliberate abrasion.

The specimens had all been removed from the safe by Dr. Oakley, who, without a word as to his own verdict, handed them to us, the jaw to Profession Le Gros Clark, the canine to me. (In exchange I gave him the chimpanzee molar I had filed down and stained at Oxford.) The examination gave us evidence in plenty that the condition of the grinding surfaces of the teeth were fully consistent with the action of deliberate abrasion, as Dr. Oakley had concluded before our arrival. On the canine could be seen without any difficulty the very scratches of the abrasive; equally obvious were the scratchings on the isolated molar from site II. The two molars in the jaw were well polished over most of their biting surface, but on some of the cusps the tell-tale scratches could be seen. The polishing itself appeared artificial.

There was much more than this to provide detailed confirmation.

As we knew from the casts, the occlusal surface of the molars was planed down to a flatness much more even than that seen in natural wear in apes' teeth. Indeed, flatness approaching that of the Piltdown molars is to be found only in aged apes with commonly the eye-tooth broken away, and at such a stage of attrition the tooth would be worn very far down. But on the 'fossils' the teeth are already flat at an early stage, altogether unusual for natural wear. In these originals we could see features completely obscured by the plaster casts. The borders of the flat, worn-down surfaces, instead of being bevelled as in natural wear, are sharp-cut, particularly on the outside edges. So too are the borders bounding the central depressed basins of the molar teeth. These edges are unnaturally sharp-cut while the floor of this basin in unworn, strange if the attrition was natural, but not surprising if a file had been applied over the tooth's surface. Another odd point is that the degree of wear of the two molars is almost identical. It is far more usual to find, at this stage of dental wear, that the first molar, erupting earlier, is more severely worn than the second.

The dental inspection tells us even more. At the points on the cusps where teeth wear away the enamel gets removed and the dentine becomes exposed. Now, in normal wear, as the dentine is softer, its level tends to be lower than that of the surrounding enamel and a little depression forms in it. But the Piltdown cusps exhibit (Pl. 4) dentine quite flat and flush with the surrounding enamel, a state of affairs explicable only by rapid artificial rubbing down of the surface. Finally, the degree of wear on the different cusps departs from the normal sequence. Instead of greater wear and exposure of dentine on the outer cusps, as is invariable with this degree of wear, there is a complete reversal, with the inner cusps the more eroded.

As for the eye-tooth, the obvious scratches on it give good grounds for attributing its really extraordinary wear—an exposure of dentine over the entire surface from side to side— to the action of an abrasion. This severe and extensive wear is not only unlike anything found normally in ape or human canines; it is, as the reader will recall, incompatible with the immaturity of the tooth. This is judged from the X-ray picture, which shows that the tooth had only very recently erupted, for the inner cavity is large and open, quite characteristic of immature teeth of all kinds. Artificial abrasion would, of course, explain all these peculiarities of wear.

Remote from the anatomists' approach is the metal-shadowing technique for the ultra-microscopic examination of surfaces. In 1950[9] Dr. David B. Scott of the National Institute of Dental Research, Bethesda (Maryland), undertook to

examine collodion replicas of the surfaces of the Piltdown
teeth, using this metal-shadowing technique which he has
developed with Wyckoff (1946). After examining replicas of
the outer facial surfaces of the molars in the Piltdown mandible, Dr. Scott reported that 'they are not readily recognizable
as ancient teeth, since they show very little evidence of post-
mortem damage'. But, in contrast, replicas of the outer facial
surfaces of the isolated molar, and of the outer facial surface
of the crown of the canine near the tip, revealed considerable
post-mortem damage. These findings correspond precisely
with the results of the present detailed re-examination of the
teeth, which have shown that the molars in the mandible
have been artificially abraded only on the occlusal surfaces,
whereas in the canine and isolated molar the facial surfaces
have also been smoothed artificially. The water-worn appearance of the isolated molar may have been produced by treatment with a weak acid as well as by the use of an abrasive.
The good state of preservation of the enamel on the facial
surfaces of the molars in the mandible indicates that they
were not in contact with an acid solution during the iron-
staining of this bone.

The new X-rays of canine and molar brought to light some
additional condemnatory pieces of evidence. In the first place,
the new pictures were needed to decide whether there was any
force in another of the initial objections which could be raised
against the 'fraud' theory. Dr. Underwood had contended that
the wear on the canine, for all its unusualness, was natural
enough. The reader will recall that in opposing Lyne and as a
supporter of Woodward he had claimed that 'secondary
dentine', as was to be expected, could be seen in the X-rays.
But our new and much clearer pictures showed no evidence

whatever of such a deposition; there was no sign of the closing up of the cavity through secondarily deposited dentine which should have been very evident with so severe a degree of attrition. A little plaque of material which had been taken to denote some secondary dentine turned out to be a small mass of some plastic material at a point where the pulp cavity had actually come extremely close to the surface. In fact this material appears to plug an opening in the apex of the cavity to the outside—a wholly unnatural state of affairs and again only understandable as a consequence of artificial abrasion. Incidentally, in the X-ray picture of the canine the shadow of the tooth's outline does not appear at all firm. This fuzziness confirms what the chemical analyses of the surface coating already indicated, that the coating was not wholly iron oxide, for in this case it would have been more distinct. As it is we know that the coating is Vandyke brown, which is partly bituminous. Dr. Claringbull even found a minute spherule of a metallic alloy embedded in the coating on the labial surface of the crown.

The X-rays served to dispose also of the last of the three serious objections to our belief in the jaw as that of modern ape (orang or chimpanzee). In the old X-ray of the jaw first published by Dr. Underwood, at a time when dental radiography had not yet reached an invariably high standard, the roots of the molars presented a quite unape-like condition. They appeared rather short and stumpy, and suggestively human. Our recent radiographs disposed of this belief completely. The apparent shortness is due to the tip of the most forward root having been broken off. This broken off piece and the ends of the other specially short roots simply did not show up in the very poor original X-ray. One should add that the inked outlines of these roots were figured in various papers, so that even the original X-rays, bad as they are, were not examined as closely as they might have been.

The treatment which the mandible has received in order to 'fossilize' it explains very reasonably the presence of cracks on the surface of the bone. This pattern is very like the stress or split-line patterns produced in modern jaw bones which have been slightly decalcified and dried, as Dr. S. L. Washburn pointed out in July 1953. The treatment probably included not only drying, but possibly immersion in dilute acid to smooth fractured edges and thus simulate the wear due to abrasion in a river bed.

The objective testimony of the morphological and radiological examination furnishes a body of evidence quite as comprehensive as does the chemical and physical. As with the latter techniques, the fragments at both sites bear many different signs, ascertained by different methods, of the tampering to which the material has been subjected.

The anatomical, like the chemical and physical, investigations had disposed of postulated objections, confirmed the suggestions provided by the casts, and given unexpected confirmation in a number of different ways. The examinations had convinced us that the jaw was not that of a form of chimpanzee, as Woodward's first critics, Waterston in this country, Miller in the United States and Ramström of Sweden,[10] had maintained, but belonged to an orang utan, a view[11] first put forward in 1927 by Frassetto and by Friederichs in 1932, and supported by Weidenreich, who all regarded it as genuinely fossil.

Of our reasons for this we may mention in particular that the height of the crown of the molar teeth and the shape of the pulp cavities seem to us quite unlike those of the chimpanzee. Friederichs advanced several good reasons for his views in his elaborate study of the detailed anatomy of the jaw.

Finally, we have been able to obtain a close similarity, anatomically and radiographically, to the Piltdown jaw in a female orang jaw by appropriate maltreatment (Pls. 4 and 5). This shows the finer points of detail on the exposed dentine of the abraded teeth very well. Particularly impressive is our artificially abraded canine. Apart from its somewhat greater size, it is almost an exact replica of the Piltdown original.

The completeness of the coherence between anatomical and chemical evidence can be easily illustrated. The independent evidence of the chemical tests is such that the extraordinary nature of the wear of the teeth cannot be other than fraudulent, since in a modern ape such characters cannot be matched; conversely, the independent anatomical evidence of the maltreatment of the teeth leads one to predict in detail (as we did) the very results of the chemical tests. When the reader recalls, in addition, that evolutionary and chronological considerations make the real existence of *Eoanthropus dawsoni* in the highest degree incredible, then the exact correspondence between the anatomical and the other tests is altogether comprehensible. The Man of Piltdown was an artifact.

Before leaving, for the time being, the skull bones of 'Piltdown man', we must refer to the serious and disturbing matter of their chromium-stained condition. This is a complicated aspect of the whole affair and a complete evaluation of the significance of the use of chromium on the Piltdown material will be attempted later.

That some of the cranial bones had been treated with bichromate of potash was well known. Sir Arthur Keith knew for a long time that some such treatment had been employed. Smith Woodward recorded that 'the colour of the pieces which were first discovered was altered a little by Mr. Dawson when he dipped them in a solution of bichromate of potash in the mistaken idea that this would harden them'. Direct chemical analysis carried out by Drs. M. H. Hey and A. A. Moss in the Department of Minerals at the British Museum (Natural History), as well as the X-ray spectrographic method of Dr. E. T. Hall in the Clarendon Laboratory, Oxford University, confirmed that all the cranial fragments seen by Smith Woodward in the spring of 1912 (before he began systematic excavations) do contain chromate; on the other hand, there is no chromate in the cranial fragments subsequently collected that summer— either in the right parietal or in the small occipital fragment found *in situ* by Smith Woodward himself. This being so, it is not to be expected that the mandible (which was excavated later and in the presence of Smith Woodward) would be chromate-stained. In fact, as shown by direct chemical analysis carried out in the Department of Minerals of the British Museum, the jaw does contain chromate. It is clear from Smith Woodward's statement about the staining of the cranial fragments of Piltdown I (which we have verified), that a chromate-staining of the jaw could hardly have been carried out without his knowledge *after* excavation. The iron- and chromate-staining of the Piltdown jaw seemed to us to be explicable only as a necessary part of the deliberate matching of the jaw of a modern ape with the mineralized cranial fragments.

In the later stages of our investigation definite evidence was obtained, and this will be presented in due course, of the fraudulent nature of the iron-staining on many specimens said to have come from the Piltdown sites. By means of an intensive chemical and crystallographic study this was found to be true of the cranial fragments of both Piltdown men.

The creation of the composite man-ape, Piltdown man, was evidently an elaborate affair; much thought and work had gone into the preparation of the fraudulent jaw and in the provision of the other items of the deception. We can discern in this elaboration the whole history of the successive discoveries, each new find adding to the whole case for the fossil man. Thus we see the discovery of an ancient gravel formation followed by the finding of a thick fossilized cranium, and this by the remarkable simian mandible, then comes the equally remarkable eye tooth and in due course the fragments of a second composite creature. And as if this was not persuasion enough, there is still the weighty ballast of the animal bones and the implements. As we are now aware beyond doubt of the spurious nature of some of these elements in the discovery, we naturally wonder about them all.

---

1. Mr. C. F. M. Fryd and Mr. A. D. Baynes-Cope (Government Chemist's Dept.), Dr. G. F. Claringbull, Dr. M. H. Hey and Mrs. A. Foster (Mineral Dept., British Museum), Dr. A. V. M. Martin (King's College, London), Dr. G. Weiler and Dr. F. B. Strauss (Oxford), Mr. S. H. U. Bowie and Dr. C. F. Davidson (Geological Survey), Dr. A. E. A. Werner and Miss R. J. Plesters (National Gallery).

2. Cooke, S. F., and Heizer, R. F., 1947, 'The Quantitative Investigation of Aboriginal Sites: Analysis of Human Bone', *Amer. J. Phys. Anthrop*, **5**, pp. 201–20.

3. The rate of decrease varies with local conditions. The fibrous substance collagen in bone and teeth from which much of the nitrogen comes is remarkably resistant and may decay quite slowly in Arctic regions and in anaerobic deposits. In the 'open' condition of the Piltdown gravel a fossil should lose its nitrogen readily.

4. Medical Research Council's Biophysics Research Unit, King's College, London.

5. By Dr. C. Bloomfield, through the courtesy of the Director of Soil Survey.

6. A bituminous compound used by jewellers for securing gems during polishing.

7. ANALYSES OF PILTDOWN SKULL

			Percentages						C.p.m.
	N	C*	CaCO₃	P₂O₅	F/P₂O₅ (×100)	Fe	CaSO₄	Cr	†
Piltdown mandible	3·9	14·5	6·5	20·0	<0·2	3	0	0·3	<0·1
Piltdown I:									
lt. fronto-parietal (av.)	1·1	6·8	3·9	18·7	0·8	6	+	1·5	0·9
lt. temporal	0·2	4·8	3·6	23·2	0·8	8	+ +	0·7	0·4
rt. parietal	1·4	5·3	3·0	19·8	0·8	5	+ +	0·0	0·1
occipital	0·3	6·8	4·5	20·8	0·7	6 + +	0·2	0·8	
Piltdown II:									
rt. frontal	1·1	4·4	1·5	14·6	0·8	10	+ +	<0·1	<0·1
occipital	0·6	3·9	2·0	13·6	0·2	9	+ +	<0·1	0·0
Compare:									
Fresh bone	4·1	14·0	5·0	25·4	0·1	<0·01	0	0	0·0
Neolithic cranium, Coldrum	1·9	6·3	13·0	23·0	1·3	<0·1	0	0	–

* Carbon in organic fraction.   † Radio-activity in net counts per minute.

8. Still another indication of the modernity of the mandible is provided by the ash content. The specific gravity of the jaw, 2·05, also differs markedly from that of the cranial pieces of Piltdown I, 2·13. (For full details of all the tests see *Bull. Brit. Mus.* (*Nat. Hist.*) *Geol. Series*, 1955, **2**, No. 6.

9. These findings by Dr. Scott are quoted verbatim from the report in the *Bull. British Museum* (*Nat. History*) *Geol.*, 1955, **2**, No. 6.

10. Ramström, M., 1919, 'Der Piltdown-Fund', *Bull. Geol. Inst. Univ. Uppsala*, **16**, pp. 261–304.

11. Frassetto, F., 1927, 'New Views on the "Dawn Man" of Piltdown (Sussex)', *Man*, **27**, p. 121. Friederichs, H. F., 1932, 'Schaedal und Unterkiefer von Piltdown (Eoanthropus dawsoni Woodward) in neuer untersuchung.' *Z. fur Anat. u Entwicklungsgeschichte*, Bd. 98, pp. 199–226. Weidenreich, F., 1943, 'The Skull of Sinanthropus pekinensis', *Palaeont. Sin.*, whole series No. 127.

5

Flint and Fauna[1]

⌗ ⌗ ⌗

The manifestly fraudulent elements in the man-ape combination called *Eoanthropus dawsoni* are the filed down molars and canine, the Vandyke brown staining of the latter and the iron-coloration of the jaw. Taken with the massive evidence of the complete incompatibility of jaw and cranium, those fabrications assure us of the enormity of the larger deception, the foisting of a spurious fossil human ancestor on to the world of palaeontology. The plot achieved its great success because it provided in the spurious fossil a self-consistent array of evidence and this fitted well with the presumed antiquity of the gravels of the Sussex Ouse; for that antiquity there was supporting testimony in the presence of palaeolithic tools and remains of animals of the earliest phase of the Ice Age.

But now with the centre-piece proved spurious, what of its appurtenances? Since the jaw is no fossil, but a recent intrusion and a deliberate one, can we help but suspect these other objects in the gravel, impressive and persuasive as the fossil animals and implements appear on first sight?

The club-like bone implement discovered in 1914 ranks next to the skull as the most remarkable of the discoveries at Piltdown. Implements of bone are well-known to have been used in the late Ice Age, for example, by men of the cave-art period. Not only is the Piltdown specimen entirely unique in its shape, but as a primitive tool, which Dawson and Woodward were confident it was, it would rank as by far the earliest ever used; in the words of the discoverers, 'their opinion was that the working and cutting of the bone were done when it was in a comparatively fresh state'.[2] Moreover, Woodward had identified the bone as one which in all likelihood had been obtained from the femur of a very early species of elephant. Judging from the other elephant and mastodon remains, such an animal would certainly have been in existence in the times of Piltdown man.

On the occasion of the reported discovery, Reginald Smith of the Department of Antiquities of the British Museum drew attention to 'the possibility of the bone having been found and whittled in recent times', and A. S. Kennard[3] likewise had doubts whether the bone could really have been cut when fresh. But no experiments in cutting bone with flint were made by the original investigators.

Dr. Oakley argued that if bones could be whittled with crude flint tools it was difficult to understand why no comparable bone work had ever been found in the known Palaeolithic industries.

The Abbe Breuil,[4] in the light of his extensive experience of bone implements and especially of the material in the cave of Peking man, where some bone pieces might possibly have been shaped by chipping (and not cutting), did not accept the Piltdown tool as the work of early man. He suggested it might have been gnawed at by a beaver.

In 1949, at the time of writing *Man the Toolmaker*,[5] Dr. Oakley came to reconsider the unique bone implement from the Piltdown gravel pit. Woodward's suggestion of an Early Ice

Age form of elephant ante-dating the presumed age of Pilt-down man as the material for the implement he found to be untenable, as the femur of one of the larger and later Middle Pleistocene elephants could have served equally well. He estab-lished also that the worked facets were very different from the cuts produced by beavers' teeth. There was none of the charac-teristic 'pairing'. Nor had the cuts been made by a flint knife, for this produced only scratchy marks. In fact, it proved impossible to cut, in the sense of whittling, a fresh or recently dried bone, which could only be worked by flaking, scraping, sawing, or grinding. Dr. Oakley inferred that the Piltdown bone was already fossilized when it was worked and that the shaping must have been done with a metal knife. Just as we were able to 'fake' the 'faked' (Pl. 4) mandible by appropriate treatment of an orang's jaw, so Dr. Oakley reproduced the features of the fossil bone (Pl. 6) implement by Whittling a fossil bone from the Swanscombe area with a steel razor and then staining with ferric chloride.

The worked flint implements or 'palaeoliths' from Barkham Manor were recognized from the start as of exceed-ingly poor workmanship and as atypical; in the words of Ray Lankester, 'unlike any known or defined industry'. Still, the crude technique, taken in conjunction with the brown or red-dish patination and their association with so early a gravel, led Reid Moir to pronounce the flints as akin to the exceedingly old specimens from Cromer. Were it not for the coloration, the Piltdown 'palaeoliths' could easily be matched with the flint debitage found at Neolithic flint chipping or mining sites; as Dawson pointed out, 'they resemble certain rude implements occasionally found on the surface of the Chalk Downs near Lewes, which are not iron-stained'.

We decided to examine the coloration of the 'palaeoliths' and of the 'eoliths' as well. The latter, it will be recalled, are present in large numbers along with Wealden iron-stone pebbles and they are of much darker brown colour.

The surface stains of a number of flints of both kinds were tested by Dr. E. T. Hall of the Clarendon Laboratory at Oxford, using his X-ray spectrographic method of analysis. This ingenious apparatus was devised for examination of archaeological specimens without doing any damage to them, and in the past year had been used extensively by Dr. Hall[6] in the study of ancient coinage, the glazes of pottery, and the composition of bronzes. It was an ideal method for both bone and flint, and its use represents yet another of the novel methods brought to bear on the Piltdown problem.

These X-ray analyses showed all the flints to be ferruginous, as expected. The 'palaeolith' flint (No. E.606), however, provided a notable exception. This flint in general shape resembles a small pointed hand-axe and was regarded by Woodward and Dawson as exhibiting the best workmanship of the flints actually recovered from the pit. It is an important discovery, because, unlike the others, which came from the heaps of gravel rubbish, this one was found *in situ* by Father Teilhard de Chardin in the middle stratum of the Piltdown gravel overlying the bed from where the mandible and one fragment of occipital bone were said to have been obtained. It thus provided firm and collateral evidence of the antiquity of the 'human' bones. But that this fragment has been deliberately stained admits of no doubt. The spectrographic analyses showed that in addition to iron the yellowish-brown stain contained appreciable traces of chromium, and this result has been confirmed by Dr. A. A. Moss of the British Museum

(Natural History), using a direct chemical method of analysis. Chromium cannot be detected in the Piltdown gravel, or in any of the naturally occurring stones and flints. It may be presumed that the flint was dipped in a solution of an iron salt and then in a bichromate solution or in chromic acid with the aim of assisting the oxidation (rusty colours of flints are due to oxides of iron).

Even the iron-staining itself looks dubious. In brown flints normally found in the Piltdown gravel the cortex is iron-stained throughout its thickness, but when a small chip was removed from this flint (E.606) for chemical analysis, the staining proved unexpectedly superficial; below its surface the cortex proved to be pure white.

The palaeoliths at Piltdown invite the gravest suspicion. The number of flints actually found at Piltdown is exceedingly small, four or five at most; in the recent extensive re-excavation by Mr. Toombs no humanly fractured flint came to light. In view of the geologically revised date, it is hardly likely that these flints can be anything as old as 'pre-Challean'. One wonders whether any worked flints occur at all in the Piltdown gravel. We shall have occasion, in due course, to return to our belief that the palaeoliths are to be explained as neolithic flints, suitably stained and planted in the gravel.

Scrutiny of the implements has thus gone well beyond the undermining of their values as archaeological documentation of the cultural activities of the Piltdown maker and therefore as testimony to his existence. The examination has brought to light new elements of fraudulent activity, in the working of the bone implement, and the iron-staining of palaeolith E.606 with the aid of a chromium compound. As the implements prove highly suspect, what of the associated animal remains? If the implements were introduced into the gravel, not only as evidence of Piltdown man's culture, but to strengthen the case for his extreme antiquity, could not the fossils have played a similar chronological role?

Thus the occurrence of the *in situ* flint E.606 *above* the *in situ* piece of occipital cranial bone and the *in situ* mandible could well suggest an antiquity as old as the more basal parts of the stratum. And this basal part on the evidence of the fossil animals found throughout the strata could be taken as indicating a 'pre-Ice Age' (or 'Upper Pliocene'[7]) dating. It is a fact that some of the Piltdown fossils (*Elephas* ('Stegodon'), *Mastodon*, and *Rhinoceros*) are undoubtedly 'Upper Pliocene' (of earlier authors) or Villafranchian, while others are not older than Middle or Upper Pleistocene. When they were first described there were firm views favouring the presence of gravels of two ages at the site. Thus, according to Keith,[8]

> Mr. Lewis Abbott . . . expressed the decisive opinion that in the Piltdown gravel two ages were represented. The lower or bottom stratum, which contained the pliocene (mammalian) remains and human bones is, in Mr. Abbott's opinion, Pliocene in date; the upper levels in which the crude Palaeolithic implements lay have been disturbed at a later time, and are to be regarded as Pleistocene in age.

A strong point for associating the dark-brown human remains with the undoubtedly older dark-coloured 'Pliocene' group was that the latter included some specimens ('Stegodon'), which, like the unrolled cranial fragments, had sharp edges. In other words, it had to be admitted that some 'Pliocene' specimens could have found their way into the Piltdown gravel without being rolled.

The idea that the dark basal gravel was wholly 'Pliocene'

appears to have been attractive to the investigators themselves, and one which they were reluctant to abandon, even though it soon appeared untenable, for the fossils recorded as occurring *in situ* at that level included, in addition to 'Stegodon', an unrolled tooth of beaver (*Castor fiber*) which was generally accepted as belonging to the later or Pleistocene group. Thus Dawson concluded:

> It is clear that this stratified gravel at Piltdown is of Pleistocene age, but that it contains in its lower stratum animal remains derived from some destroyed Pliocene deposit probably situated not far away, and consisting of worn and broken fragments.[9]

But in a later paper he wrote: 'We cannot resist the conclusion that the third or dark bed is in the main composed of Pliocene drift . . .'[10]

That the Villafranchian fossils were fraudulently introduced, and with the very probable aim of suggesting that 'Piltdown man' dated from Pliocene times, emerges from the investigations of this material.

Between 1911 and 1914 eighteen fossil mammalian bones and teeth were found at, or in the immediate vicinity of, the Piltdown skull site. Four were recorded as having been found *in situ* in or below the chocolate-coloured basal gravel, two on the surface of the adjoining field and the remainder on the spoil heaps at the edge of the small pit (Pl. 2). When one considers the thinness of the gravel (average thickness 18 inches) and the small size of the pit (less than 50 × 10 yards in area), this was a remarkable yield, particularly in view of the extreme rarity of fossils in the Pleistocene river gravels of Sussex. It appears even more remarkable when one bears in mind that the gravel and overlying loam are porous deposits now completely decalcified. And despite the heterogeneity of the fossils, ranging from a heavily rolled enamel cap of a mastodon tooth to a large piece of deer antler with almost undamaged surface, they were not scattered through the gravel, but came from one or possibly two pockets. Nor, as has been recorded, could Woodward[11] in later years or Toombs[12] find any single further fragment.

The singularity of the occurrence of 'Pliocene' mammalian remains at Piltdown can be gauged from the fact that, with one exception,[13] none has been found elsewhere in southern England outside East Anglia. In the light of the more recent geological information of the Piltdown gravel obtained by Mr. Edmunds, the occurrence of this Villafranchian pocket at Piltdown becomes quite incredible. When Edmunds demonstrated that the gravel is part of a terrace 50 feet above the Ouse, it seemed that the only way of accounting for a pocket of Villafranchian fossils in that gravel was to presume that a block of indurated, fossiliferous sand of that age had worked its way down from a higher level and had disintegrated on the Ouse flood-plain in 50-feet terrace times. In that case, the association of this derived Villafranchian fauna with the remains of beaver and the human cranial fragments, seemingly contemporary with the 50-foot terrace deposits, would have been mere coincidence. But the 'coincidence' was repeated, for a rhinoceros tooth similar in preservation to the tooth of *Rhinoceros* cf. *etruscus* from the main site was reported as having been found at the Sheffield Park site, close by cranial bones and a molar tooth purporting to represent a second specimen of '*Eoanthropus*'.

The molar tooth referred to as 'Stegodon'[14] has been closely studied by Dr. Oakley, who finds it very remarkable that four pieces (probably representing two molars) should be recorded at Piltdown. Woodward himself had pointed out that such specimens had not hitherto been found in Western Europe, that they were of a type considerably earlier even than the known elephants of the Upper Pliocene and were essentially of the type known from the Siwalik formation from India. In England, the two elephant teeth from the Red Crag of Suffolk are neither as primitive as the Piltdown specimens. The latter in their reddish colour resemble the Red Crag fossils; and in their colour they also closely match the Piltdown mandible and cranial fragments. It was this similarity in colour which led some investigators to conclude (as Woodward and Dawson were inclined to do) that the Piltdown skull and mandible were indeed of Villafranchian age.[15]

Thus Dr. Oakley inferred that if the 'planifrons' molar fragments were fraudulently introduced at Piltdown it is probable that they were obtained from some foreign source (since they cannot be duplicated exactly in Red Crag collections) and artificially stained to match the Piltdown cranial bones and mandible. We know that the Piltdown mandible was given a rich mahogany colour by a process which involved the use of a chromium compound, and it is only too likely that the same method would have been applied to imported 'planifrons' teeth, since (in our experience) such fossils rarely have that precise colour except in the Red Crag. Samples of the iron-stained dentine of the two critical specimens from Piltdown (E.596 and E.620) were submitted to the Department of the Government Chemist, where they were analysed spectrographically by Mr. H. L. Bolton, who reported that they do in fact contain, as we expected, significant traces of chromium (0.3 and 0.1 per cent. respectively).

If any doubt had remained that these pieces of 'planifrons' molars were of foreign origin, it would have been dispelled by a consideration of their radio-activity. In the hope of tracing their origin, Dr. Oakley submitted to Mr. Bowie and Dr. Davidson[16] for radio-activity tests a series of mammalian teeth from the main Villafranchian fossil localities. The results obtained, summarized below, confirm the conclusion that the Piltdown specimens were not obtained from an English deposit. It can be taken that a specimen with a radio-activity of about 200 counts per minute (c.p.m.) is extremely unlikely to have come from a deposit in which other samples (chosen at random) show a radio-activity falling below 30 c.p.m. The (maximum) counts of the three Piltdown pieces of molar were respectively 355, 203 and 175 per minute. Fossils from nearly a dozen Villafranchian localities in Europe and in Pakistan all failed to show radio-activity in excess of 28 c.p.m. Six specimens from East Anglia were below 15 c.p.m. But from one locality, Ichkeul in Tunisia, a tooth was tested[17] which proved to have a radio-activity (max. 195 c.p.m.) practically identical with the Piltdown specimens.[18] Ichkeul, not far from the port of Bizerta, is a place where fossil mammalian remains, in particular *Elephas* cf. *planifrons*, are very abundant.[19]

These teeth of *Elephas* from Tunisia can be given an iron and chromium treatment to match the coloration of the Piltdown specimens. The presence of the tell-tale chromium in these specimens gives good reason to believe that the iron-staining was likewise deliberate.

The remains of the other Villafranchian (or so-called Pliocene) species, *Mastodon* and *Rhinoceros* (the latter represented by a find also at Sheffield Park), like 'Stegodon' contain fluorine in high concentration (1·9–3·1 per cent.) but there is no evidence of artificial staining. Nor would these specimens require any treatment. In their morphology, colour, mineralization, radio-activity, and fluorine content they are indistinguishable from specimens of the Red Crag of East Anglia from where in all probability they originally came.

The fluorine (and radio-activity) results clearly supported the generally accepted division, as judged by colour, mineralization, and morphological characters of the Piltdown fauna into an older high fluorine group and a more recent group with fluorine values ranging rather widely, from 0·1 to 1·5 per cent. Into this latter group, along with beaver, deer, and horse, as we know, went the *Eoanthropus* material, with all the consequences we have described.

Very unexpected too was the low fluorine content of the *Hippopotamus* teeth, especially of the molar, for this had always been placed with the older, Villafranchian group because of its dark colour and for the excellent palaeontological reason that the form is always an ancient one in Britain. The hippo molar which had once aided *Eoanthropus* to secure a foothold in the older Villafranchian level, continued even after the first fluorine tests by Oakley and Hoskins to assure some measure of antiquity to the Piltdown skull (and jaw)—but only as long as the hippopotamus tooth with the same low fluorine content was accepted as a genuine local fossil. For it could be argued that the ground water at Piltdown had been exceptionally deficient in fluorine since the Ice Age (the fluorine content of the Piltdown skull is lower than that of fossils from any of the Pleistocene gravels in Britain). When evidence came to light that the hippo tooth was also a fraudulent introduction, the provisional dating of the Piltdown brain-case as Late Pleistocene[20] lost its foundation and the status of the brain-case itself as a genuine local fossil was brought into question.

The enamel of the hippo molar is colourless, but its dentine containing only 0·05 per cent. fluorine is stained brownish-black throughout. Its state of preservation is altogether different from that of the 'contemporaneous' beaver teeth. As with 'Stegodon', the suspicion of artificial staining was at once confirmed by the finding in the dentine of 1 per cent. chromium, and the iron-staining, as will be shown later, is also artificial. Yet the molar is without doubt a fossil—it has a very low organic (nitrogen) content and a correspondingly low ash content. The combination, in it, of a low fluorine and a low organic content suggested to Oakley that the specimen had come from a limestone cave deposit in which bones and teeth absorb fluorine only very slowly, but their organic content decreases at a normal rate. The 'Piltdown' molar can be matched very closely both in chemical composition and dental morphology with that from the cave deposits of Mediterranean islands, for example Malta. Hippopotamus teeth from Maltese caves are creamy white in colour but by staining with ferrous sulphate the Piltdown dark brown colour so persuasive of Red Crag or 'Pliocene' affinity can be closely reproduced.

The second hippo specimen—the pre-molar—claims a special place as the first fossil from Piltdown shown to Woodward. Its fluorine content shows it to be derived from a different source from that of the molar. Like that tooth, this is dark-coloured and contains the incriminating chromium.

The beaver incisor and piece of mandible, the metatarsal of the deer—all are artificially stained, and Dr. Oakley concludes they came from relatively recent deposits. He has also come upon one fabrication quite bizarre, though perhaps not unduly so by Piltdown standards. It is furnished by the fragments of cream-coloured bone said to have been found *in situ* in clay at the base of the gravel deposit and regarded by Dawson and Woodward[21] as indicating the source of the worked slab of elephant femur—Piltdown Man's famous 'cricket-bat' or 'club'.

> One of those pieces is still embedded in a lump of loam, adhering to the middle of a slab of ironstone. The loamy matrix shows every indication of being faked. It contains small scattered pebbles set at various angles, and it shows cracks and cells of burst air-bubbles such as those which are liable to appear if loamy matter is worked into a paste and then allowed to set.[22]

'Of the eighteen specimens of fossil mammals recorded from the Piltdown gravel by Dawson and Woodward', writes Oakley, 'ten are unquestionably frauds, and there are strong grounds for believing that this is also true of the remainder.'[23]

1. This chapter and the next are based on an extensive paraphrase of Dr. Oakley's contributions in the second Piltdown report (*Bull. Brit. Mus. (Nat. Hist.) Geol.*, 1955, **2**, No. 6).

2. Dawson, C., and Woodward, A. S., 1915, 'On a Bone Implement from Piltdown, Sussex', *Quart. J. Geol. Soc.*, **71**, 144.

3. Dawson and Woodward, 1915, op. cit., Discussion.

4. Breuil, H., 1938, 'The Use of Bone Implements in the Old Palaeolithic', *Antiquity*, **12**, pp. 56–67.

5. Oakley, K. P., 1949, *Man the Toolmaker*, 98 pp., Brit. Mus. (Nat. Hist.), London.

6. Hall, E. T., 1953. 'Analysis of Archaeological Specimens: A New Method', *The Times Science Review*, No. 9, p. 13.

7. Now called 'Villafranchian' and regarded as Lower Pleistocene.

8. Keith, A., op. cit., p. 508.

9. Dawson and Woodward, 1913, p. 123.

10. Dawson and Woodward, 1914, 'Supplementary Note on the Discovery of a Palaeolithic Skull and Mandible at Piltdown (Sussex)', *Quart. J. Geol. Soc. Lond.*, **70**, pp. 82–93.

11. Woodward, A. S., 1917, op. cit., pp. 12–13.

12. Toombs, op. cit.

13. This is a very dubious specimen of extinct bear from Portslade.

14. More accurately designated as *Elephas* cf. *planifrons*.

15. Hopwood, A. T., 1935, 'Fossil Elephants and Man', *Proc. Geol. Assoc.*, **46**, pp. 46–60.

16. Atomic Energy Division, H.M. Geological Survey.

17. Kindly supplied by Professor C. Arambourg.

18. There is also close agreement in the fluorine content of these specimens.

19. Arambourg, C., and Arnould, M., 1950, 'Note sur les fouilles paléontologiques executées en 1947–48 et 1949 dans le gisement Villafranchian de la Garaet Ichkeul, *Bull. Soc. Sci. Nat. Tunisie*, II, fasc. 3–4, pp. 149–57.

20. Oakley and Hoskins, 1950, op. cit.

21. 1915, op. cit.

22. Oakley, K. P., in *Bull. Brit. Museum (Nat. History)*, 1955.

23. It will be as well to deny categorically and absolutely a statement issued in June 1954 through *Science Service*, Washington, D.C., to the effect that 'when the Piltdown hoax was exposed at the meeting of the Geological Society of London in November 1953, it precipitated a violent discussion. . . . The meeting soon broke up into a series of fish-fights. . . . The fracas resulted in the expulsion of several members. . . .' There was in fact no general discussion and no disturbance of any kind at any of the meetings at which Piltdown was discussed.

6

The Full Extent

¤ ¤ ¤

Then the perilous path was planted
And a river and a spring
On every cliff and tomb,
And on the bleached bones
Red clay brought forth.
          Blake: *The Marriage of Heaven and Hell.*

Almost any single one of the techniques employed in the investigations suffices to reveal the elaborateness of the deception which was perpetrated at Piltdown. The anatomical examination, the tests for fluorine and nitrogen bear particularly good witness to this; even the radio-activity results taken alone, led the physicists to remark on the 'great range of activity shown by specimens from this one little site'; 'it is difficult to avoid the conclusion that the different bones in the Piltdown assemblage have had very different geological and chemical histories'.

We have merely to take account of the stained condition of the whole assemblage, to realize the thoroughness of the fraud. From the Vandyke brown colour of the unnaturally abraded canine we infer with certainty that it was deliberately 'planted'. The superficiality of the iron impregnation, combined with the chromium, tells as much as regards the orang jaw. And it is this iron-staining which finally shows that the rest, human and animal, was without doubt, all 'planted'. The iron-staining has two peculiar features. It seems probable that ferric ammonium sulphate (iron alum) was the salt employed. This salt is slightly acid. The peculiarity of this salt (and, indeed, of any acid sulphate) is that in bone which contains little organic matter such as the cranium of Piltdown I, or Piltdown II, the beaver bones and hippo teeth, it brings about a detectable change in the crystal structure of the bone. In the apatite in which the calcium of the bone is held, the phosphate is replaced by sulphate to form gypsum. This change is quite unnatural, for neither gypsum nor sufficient sulphate occur in the gravels at Piltdown to bring it about. So the iron-sulphate-staining is an integral part of the forger's necessary technique. He also used chromium compounds to aid the iron-staining probably because he thought it would assist the production of iron oxide. Chromium compounds are oxidizing.

The basic strategy underlying the Piltdown series of forgeries now seems reasonably clear. Two main elements in the plan taken together explain nearly all the features of the affair quite satisfactorily.

In the first place there is the consistent intention to establish beyond anyone's questioning the occurrence of a Pliocene or Red Crag layer in the Sussex Downs. The finding of such a layer would be no real surprise to many. It would accord with the expectations long held by the 'Ightham Circle'[1] and other Pleistocene palaeontologists, who had persistently searched the Weald and the Downs for just such a Pliocene deposit. The Piltdown gravel yielded up exactly what was sought. In it, recognizable to all, were the teeth of rhinoceros and mammoth specimens indistinguishable in every way from that of the Red Crag. They are genuine enough. They have not been stained either with iron sulphate or chromium. But their chemical composition (as well as their morphology) is identical with other Red Crag specimens so as to leave little doubt that East Anglia is their real place of origin. These objects with their rich red mahogany colour provide the pattern for all the other 'finds' intended to bolster up this Pliocene horizon. Accordingly, the 'planifrons' elephant tooth which comes from nowhere in Europe has been given an appropriate iron colour and contains chromium; the hippo teeth, again extraneous, have been given a very natural reddish gravel colour by the iron sulphate and chromium method, and the fauna of the 'later' layers has also been treated.

The 'human' remains have thus emerged from a gravel bed containing the 'Pliocene' looking animal bones and it is highly probable that the plan was to secure this dating for the jaw and cranium—the long-awaited, often-heralded Pliocene Man.[2] For those archaeologists who set store by them there were provided the abundant so-called 'eoliths' to testify to the Pliocene date of their supposed maker. Some enthusiasts of the 'eolith' theory had it both ways, for Piltdown Man could be invoked to prove the Pliocene date of the eoliths.[3] Whether the intention was to endow Piltdown Man with a kit containing eoliths and the fraudulent and suitably stained palaeoliths as well, we cannot say. In the event, the apparent position of the flint tools above the deepest layer, combined with the presence of fossil fauna such as beaver and deer, led to the general conclusion that Piltdown man, though a Dawn Man, was somewhat younger than *Elephas* or *Rhinoceros*. Yet, so convincing is the colour and the apparent mineralization that some authorities could continue to argue with some justice that the human remains were quite likely attributable to the lowest layer. As we have noted already, even differences in 'rolling' could not be urged against this view. Woodward was quite inclined to this even after 1926, and he thought that the bone implement fashioned from what he considered to be a Pliocene elephant, served as strong evidence as it seemed to have come from the lowest level of all in the gravels.

Whether he was pre-Pleistocene or not, Piltdown man still retained in the scientific world his antiquity as an early Pleistocene man equipped with appropriate tools of the period. The forger's intention may have gone slightly astray in not attaining the most ancient heritage possible for his brain-child, but the experts' verdict served well enough for forty years. As we now know, there is nothing at all to sustain the antiquity of the gravels, and we can dismiss the fauna as an assortment of importations from at least three different sources.

The second main theme running through the whole affair is of course the 'build-up' of the man-ape combination. The cranial pieces arrived first accompanied by their 'certificates' of antiquity, the hippo teeth and pieces of 'Stegodon' (*Elephas*) molar. All those delivered to Smith Woodward in London were chromium- and iron-stained. Then, closely associated with further pieces of cranium, came the famous jaw, along with further pieces of datable 'Red Crag' remains. The rhinoceros needed no staining; and though the cranial pieces had now no chromium, as had the 'Stegodon', all had the tell-tale sulphate. The canine, we know, consolidated the disputed position of the newly-recognized *Eoanthropus*, and the position was secured when Piltdown II arrived, again artificially stained with sulphate and chromium.

The significance of the finds at the second Piltdown site should be clearly understood. The occipital fragment undoubtedly comes from a second individual, but it does not belong to the frontal bone with which it was obtained. The occipital fragment is not remarkable in thickness nor in other morphological features, but its neat rectangular outline strongly suggests that it has been trimmed to that shape, as does also the rectangular shape of the frontal. The frontal bone is in texture and thickness very similar to the bones of the find of Piltdown I. Anatomically, as Keith and others recognized years ago, it can well fit into a missing part of the first skull. It is true that the Piltdown skull can be matched among some recent crania, but such skulls are undoubtedly rare, so that to find two different skulls in the same condition would be very unlikely. Indeed, it is quite possible that the peculiar thickening of the Piltdown I and the frontal of Piltdown II is the result of some pathological process. This was the view[4] of the late Professor S. G. Shattock, Pathologist of the Royal College of Surgeons, and an authority on bone pathology. Chemically, it is in fact part of the first skull. The feature in the composition of the bones least likely to be affected by the iron sulphate and chromate treatment (which Piltdown II, like Piltdown I, has received) is the fluorine-phosphate ratio. In this the frontal of II agrees with the bones of skull I and not with the occipital which accompanied it. It is a piece retained by the forger for many years. The second Piltdown Man purports to be a complete duplication of the first ape-man combination, by alignment of the isolated molar tooth with the cranial pieces, and furnished with the necessary Pliocene dating specimen—a piece of Red Crag rhinoceros tooth.

The great success of the Piltdown hoax came from the clear

conception on the part of the perpetrator that a man-ape of the right age appearing in the hitherto unknown gravel had a good chance of deceiving the palaeontological world. He planned and worked to admirable effect to provide a man-ape at Barkham Manor which would stand up to recurrent criticism, and to furnish him with an antique milieu adequately stocked with the appropriate animal fauna and the man-ape's tools. He was able to stage a second, if paler, version at Sheffield Park.

The cleverness of the hoaxer needs no stressing. It can be illustrated by the matter of the grains of sand packed in the pulp cavity of the eye tooth. Seen in an X-ray (as they were first in 1913) there are nineteen of these grains, most of them opaque. The largest grain plugs the opening to the cavity, and the whole gives a strong impression of fossilization. But careful consideration reveals a different story. The grains are loose, they rattle and are not consolidated, as might perhaps be expected. Some were extracted and proved to be pellets of limonitic iron-stone such as occur in the sand fraction of the Piltdown gravel. But very striking is the virtual absence of the fine sandy material below 1 millimetre in diameter which is found in a frequency of 30 per cent. in the sand at Piltdown. If the filling had been natural, as by silting in, this finer material would have been present. We have found it easy to fill and plug our own 'faked canine' in just the same way.

Well-executed and resourceful as the whole plan now appears, the existence of serious initial weaknesses should not be overlooked. Perfection is not easily attained even in this curious technology. The canine tooth, perhaps the boldest of the products, is probably at the same time the weakest. The use of bituminous Vandyke brown rather than an iron salt, so unlike the colouring used on the other objects, was a fatal flaw.

But in all fairness it should be admitted that the use of this paint was probably forced on to the forger, for the staining of fresh teeth by means of iron with or without chromium is, in our experience, by no means easy. To get a good fossil appearance, there was probably no other recourse but the paint. This tooth declares the work of the forger again in the unnatural condition in which the occlusal surface has been left as the result of its excessive abrasion. The canal has been opened up at one point and become blocked by the paint.

The unnecessary use of a chromium compound to assist the iron-staining of one of the flint tools must surely be judged a glaring miscalculation or oversight. The fabricated bone implement too seems a more than ordinary risk to have taken, for its complete unexpectedness and uniqueness were bound to attract rather special scrutiny. It was fortunate for the forger that Reginald Smith, who voiced an unmistakable scepticism, did not himself proceed to do what he suggested—namely, to discover by experiment whether fresh bone could possibly be whittled by any kind of flint tool.

Piltdown II seems in retrospect rather hazardous. The obvious contrast of the occipital fragment to the frontal accompanying it, and its equally obvious resemblance to the pieces of frontal from the original site, did not escape some observers early on. One has the impression that the very severe criticism by Waterston and Miller of Woodward's *Eoanthropus* coming so soon must have driven the hoaxer to provide at Sheffield Park this weaker imitation of Barkham Manor, just as the canine was produced to satisfy critics at an earlier stage.

Admiration of the cleverness of the whole design must therefore be tempered by a realization of the good luck which saved the forger from immediate exposure on several occasions.

An escape even more striking than those already mentioned was the failure of the investigators to secure from the Uckfield Public Analyst a chemical examination of the mandible for comparison with that which he had already carried out on the cranium. In the latter he had reported a complete absence of organic matter. We know that the jaw contains as much organic matter as fresh bone, so that the analyst could not have failed to find a very striking discrepancy between the mandible and the cranium. Had he done so, the exact circumstances leading to the sensational rejection of the Moulin Quignon jaw in 1863 by Busk and Prestwich would have been re-enacted. It was the high nitrogen content of that bone which convinced the English investigators that the jaw (and with it an isolated human tooth) were forgeries which, along with a number of palaeoliths, had been foisted by the workmen at Abbeville on Boucher de Perthes.

----

1. See later, pp. 96 and 104.

2. The Foxhall Man, among others had been claimed as a representative of 'Pliocene Man' but the indications were against it as everything pointed to a natural burial.

3. For example, by H. Morris (see later, p. 160).

4. Shattock, S. G., 1913, 'Morbid Thickening of the Calvaria; and the Reconstruction of Bone once Abnormal: a Pathological Basis for the Study of the Thickening observed in Certain Pleistocene Crania', Report XVIIA, *Internat. Med. Cong.*, Sect. III (General Pathology), pp. 3–46.

7

The Principals and Their Part

◻ ◻ ◻

The objective evidence for the deception at Piltdown was overwhelming. The frauds extended to every aspect of the discovery—geological, archaeological, anatomical, and chemical—so that proof could be adduced three or four times over. Moreover, every time a new line of investigation was applied, it confirmed, as we have seen, what all previous evidence had established. The two Piltdown 'men' were forgeries, the tools were falsifications, the animal remains had been planted. The skill of the deception should not be underestimated, and it is not at all difficult to understand why forty years should have elapsed before the exposure; for it needed all the new discoveries of palaeontology to arouse suspicion, and completely new chemical and X-ray techniques to prove the suspicion justified.

Professor Le Gros Clark, Dr. Oakley, and I wrote in our report[1] that 'Those who took part in the excavation at Piltdown had been the victims of an elaborate and inexplicable deception'. Inexplicable, indeed, for the principals were known to us as men of acknowledged distinction and highly experienced in palaeontological investigation. Woodward, in 1912, was a man of established reputation. Dawson enjoyed a solid esteem. Teilhard de Chardin was, of course, only at the beginning of his palaeontological career. Knowing their place in the world of science, we felt sure that these investigators, whose integrity there was not the slightest reason to question, had been victims—like the scientific world at large—of the deception.

Arthur Smith Woodward (who was of an age with Dawson) at the time of the discovery had been Keeper of the Department of Geology at the British Museum since 1901, the year of his election to the Royal Society, and had scores of papers of very great merit to his credit. His work on fossil reptiles and fishes was on a monumental scale, and he had also made discoveries in mammalian palaeontology. He was without doubt the leading authority in his own field. His position was abundantly recognized by many awards and by appointment to many high offices—for example, Secretary, and in the Piltdown years successively Vice-President and President, of the Geological Society.

Woodward had started in the Museum as an Assistant, entering by competitive examination at the age of eighteen, and educating himself by attending evening classes at King's College while working at the Museum, and had very soon shown his ability as a research palaeontologist. After only five years he was entrusted with the preparation of an exhaustive *Catalogue of Fossil Fishes*, the evolutionary interpretation of which he greatly clarified. He was careful and painstaking in the highest degree, with intense powers of observation and perseverance. As Keeper he maintained a multitude of personal contacts in the world of geology, professional and amateur. From his correspondence one is amazed at the immensity of the material which kept flowing to him at the British Museum for his observation and criticism, and to which he attended without abatement even in the Piltdown years. Practically every year before his retirement he travelled abroad to study material in Continental museums and to visit interesting sites. Though his main preoccupation was with fossil fish and reptiles, through his travels he became acquainted at first hand with the Java fossils of Dubois at Leyden, with the Mauer jaw at Heidelberg, and with the various specimens of Neanderthal and other Pleistocene men.

When Dawson wrote to him on 14 February 1912 to tell him that he had 'come across a very old Pleistocene bed overlying the Hastings beds between Uckfield and Crowborough which I think is going to be interesting' with 'part of a thick human skull in it', he was addressing a man of great experience and judgement. Some may have thought that Dawson should have brought his discovery to the attention not of Woodward, but of Arthur Keith, who was then Conservator at the Royal College of Surgeons, and indeed Keith himself thought so. Yet it was perfectly natural for Dawson to turn to Woodward with his experience of mammalian palaeontology, the more so as they had known each other for over twenty-five years and had worked together from time to time. One such occasion was as far back as 1891, when Woodward described the first tooth of a new Wealden mammal[2]—*Plagiaulax dawsoni*—obtained by Dawson from one of the fine, pebbly bone-beds of the Wealden series. Dawson was in consultation with Woodward again in 1909 in connection with some more teeth of these Wealden early mammals, including, besides two more *Plagiaulax*, a new form called *Dipriodon*. They were, and remained, on terms of fairly close friendship. Dawson called on the Woodwards (his signature appears on Lady Smith Woodward's famous embroidered table-cloth,[3] along with the signatures of hundreds of other notabilities whom the

Woodwards had entertained). He referred in 1915 to Smith Woodward as 'my old friend'.[4]

That Dawson enjoyed Woodward's full confidence and high esteem is quite certain from the obituaries Smith Woodward wrote of his collaborator,[5] and from the description of the man in *The Earliest Englishman*. The respect for Dawson's attainments was shared equally by Arthur Keith.[6]

In the geological world of 1913 Dawson had a secure and increasing reputation, and we know for certain that his geological colleagues (including many Fellows of the Royal Society) were anxious to give him the highest recognition for his many years' work in geology.

Sir Arthur Keith has written of Dawson as the lawyer-antiquarian, the exemplar of the English country amateur, a description which is well known to anthropologists at large from the book[7] by the American anthropologist, Hooton of Harvard, a staunch supporter of *E. dawsoni*. But it would be entirely wrong to think of Dawson as an untrained amateur. From his schooldays onwards his interest in geology and archaeology had been unremitting. While still a schoolboy at the Royal Academy, Gosport, he had begun to search the Weald for fossil reptiles under the tutelage of S. H. Beckles, F.R.S., then in his last years at St. Leonards-on-Sea. So successful was he that he was able to present to the British Museum an impressive collection of Wealden fossils which, along with those of his old friend, Beckles, he put in order and catalogued. By the early age of twenty-one his work in geology brought him a Fellowship of the Geological Society, to which he was admitted on the same day as Arthur Smith Woodward in 1885. He was accepted as an honorary collector for the British Museum for over thirty years. The Dawson Collection at

South Kensington, to which he constantly added, contains some highly important specimens. He was responsible for finding three new species of iguanodon, one of which was named *Iguanodon dawsoni*, as well as other dinosaurs and the Wealden mammal, *Plagiaulax dawsoni*.

Dawson's father was a barrister-at-law living at St. Leonards, and, after leaving the Royal Academy, Gosport, the son was articled to Langhams, a firm of solicitors in Hastings, in 1880. Dawson spent some years in London at the head office of the firm, then went in 1890 to a branch at Uckfield, eight miles from Lewes, and became a partner. His professional career was a successful one, and in the civic affairs of Uckfield he played a prominent part, being Clerk to the Magistrates for the Uckfield Petty Sessional Division and Clerk to the Urban Council for many years. He filled many public and professional offices, including the Stewardships of the Manors of Hetherall and Camois, and in 1898 of Barkham. Nevertheless, Dawson's heart did not seem really to be in legal activities and after 1905, the year of his marriage, when Mr. Hart came into partnership with him, he spent a great part of his time pursuing his collector's interests.

In 1911, on the occasion of Woodward's report to the Geological Society of the finding of further specimens of these very early mammals, *Plagiaulax* and *Dipriodon*, Dr. H. Woodward (the namesake whom Smith Woodward had succeeded as Keeper) paid Dawson high praise for the acumen which underlay these particular discoveries. His initial success in 1891 came through intense searching for these minute teeth in thin geological strata near Hastings which he had rightly recognized, from the pioneer work of Mantell and of Marsh (the first to discover them in Wyoming), as a likely formation.

Before then there had been no trace of these particular Meso-zoic mammalia, despite the extent of the Cretaceous forma-tions in south-east England and western Europe. To this key problem of mammalian origins Dawson had by his persistent and successful searching made a contribution to add to the accomplishments of men like Owen, Boyd Dawkins, Marsh, and Cope.

That Dawson's geological abilities were of a high order is clear, and not only from his Wealden palaeontology. He wrote on 'Dene Holes', which he diagnosed as ancient mines. He recognized, and exhibited to the Geological Society, zinc-blende from the Wealden and Purbeck beds, and in 1898 made an important discovery of natural gas at Heathfield (used there to light the hotel and station for many years).

Apart from his attainments in geology, Dawson was known to us also as an archaeologist, a Fellow of the Society of Anti-quaries. His election in 1895, like his election to the Geo-logical Society, was a notable distinction for a young amateur investigator. At the time of Piltdown Dawson had already made those contributions on which rested his reputation as antiquarian. He had become known as an authority on old iron work (Straker[8] gives him special mention along with the older authority, William Topley, for his writings on the Sussex iron industry), and his large two-volume *History of Hastings Castle*[9] had become a standard work. On these topics and other matters of archaeological interest, Dawson gave communica-tions and papers to the Society of Antiquaries of London, and to the Sussex Archaeological Society of Lewes, which he had joined in 1892.

By the time he came to excavate at Piltdown Dawson could claim a not inconsiderable experience of practical field work.

In the early 1890s his study of Hastings Castle necessitated some digging and clearing of that site, and in this work he was helped by his friend, John Lewis. In 1892 we find him excavating in the Lavant caves, again with John Lewis, at the request of the Sussex Archaeological Society. In the early years of the new century he was active at the Roman camp at Pevensey and in 1906 he took part with Mr. J. E. Ray of St. Leonards in the disinterment and examination of two Iron Age skeletons near Eastbourne.

Thus the two men who undertook investigations at Piltdown were, in partnership, well-equipped to cope with the geology, palaeontology, and archaeology of that complicated assemblage.

The two men, Woodward and Dawson, provided a complete contrast in temperament and physique. Woodward's was an ascetic-looking figure, rather stern and reserved in his everyday dealings and not an easy man to get to know intimately. Keith,[10] for one, found Woodward not at all easy to get on with ('proud and cold') in the early days, but in time came to admire and like him. But for all his aloofness, Woodward retained the respect and friendship of his colleagues. His essential quality is revealed in the confidence reposed in him by the Royal Society in appointing him Chairman of the Scientific Relief Committee, which had to deal with personal problems of great delicacy. Dawson was a physically big man, energetic and genial. Woodward wrote of him as a delightful colleague in scientific research, and Keith spoke in much the same terms. Teilhard found him 'methodical and enthusiastic'.

When Woodward heard of the finds, of the 'part of a human skull which will rival *Homo Heidelbergensis*', in the letter of February 1912, he was anxious to go down to Sussex as soon as possible to see the fossil material and gravels. As the discovery seemed likely to prove of the highest importance, he counselled discretion on Dawson. Naturally enough, he did not want any premature announcements about the material, however promising the pieces of cranium and mammalian teeth so far found in the Early Ice Age gravel might be. Dawson wrote on 28 March:

> I will of course take care that no one sees the piece of skull who has any knowledge of the subject and leave it to you. On second thoughts,[11] I have decided to wait until you and I can go over by ourselves to look at the bed of gravel. It is not very far to walk from Uckfield and it will do us good!

On 26 March Woodward received from Dawson the hippopotamus tooth by post, with the request, 'will you kindly identify enclosed for me? I think the larger one is hippo.' He returned it with his report confirming that it was a 'premolar of hippopotamus' along with a piece of sandstone.

It was some months before Woodward actually saw the cranial fragments or inspected the gravel. A visit to Piltdown in March proved unsuitable; the weather was against it. 'At present the roads leading to it are impassable and excavation is out of the question', wrote Dawson on the 24th. In April the Woodwards went abroad to study dinosaurs in Germany. Dawson was busy searching out the further extension of gravel beds near the original site.[12] At length, on 23 May, Dawson wrote that he would bring the precious relics along the next day. 'Some time tomorrow (Friday), probably after lunch, I will bring the piece of skull and a few odds and ends found with it, or near it, in the gravel bed.' Of this visit Dawson wrote subsequently: 'I produced my find with the remark,

"How's that for Heidelberg?" [13] The long-planned visit to Piltdown followed soon after. The arrangements were made as unobtrusively as possible. When Woodward finally went down, on 2 June, Teilhard de Chardin made a third. Dawson had written on 27 May: 'Next Saturday (2 June) I am going to have a dig at the gravel bed and Fr. Teilhard will be with me. He is quite safe. Will you be able to join us?'

Excavations thus began that very first week-end. An aged labourer, named Venus Hargreaves (who died in 1917), was the only hired help. He worked under close supervision, as the pit was small and every spadeful had to be sieved. Excavation was in all probability done only on the occasions when Dawson and Woodward could get down to Barkham Manor. That first season they worked for the most part at week-ends only, Woodward coming down from London and staying near Piltdown.[14] The jaw seems to have come to light in that same June, if we surmise correctly from a letter of Dawson's to Woodward on 30 June. Here he discusses the chin region of the Ice Age Cheddah [sic] skull, asserting that it 'is nearly straight and massive'—a point of direct relevance to the problem of the missing chin region of the apparently even older Piltdown man. In 1913 the Woodward family came down for the holidays and took a cottage close to Barkham Manor. The cost of the excavations and his own expenses were borne by Woodward personally and not by the Museum.

The youngest of the principal figures at Piltdown was the Jesuit priest, Teilhard de Chardin. He had come to Hastings in 1908, after a period in Egypt and before that in Jersey. He knew about Piltdown before Woodward heard of Dawson's discovery. Accompanied by his older Jesuit friend, Felix Pelletier, Teilhard had met Dawson in a quarry outside

Hastings. These were the quarries which Dawson knew so well from his early boyhood and where he had obtained the many fossil reptiles which by 1884, as we know, were numerous enough to form the Dawson Collection in the British Museum. Dawson was still keeping up these searches, and from time to time added to the Museum's collection. Teilhard, thirty years old at the time, was a comparative newcomer to geology, which had attracted him some years previously in Egypt, for he started his academic career as a Lecturer in Physics and Chemistry in Cairo during the years 1906–8 at the College de la Sainte Famille. He was soon admitted, in 1912, to membership of the Société Géologique de France. He was a student of theology at the Jesuit College at Hastings from 1908, and his palaeontological enthusiasm was greatly encouraged by Dawson's help and generosity. Dawson had long been accustomed to tip the quarrymen for any likely-looking specimens. The workmen did not welcome the intrusion of the two priests, who did their own collecting and were not disposed to call on their services. They told Dawson about the intruders, but Dawson put it right and asked the workmen to look out for material, for which he would see they were rewarded. 'Thus began Mr. Dawson's friendship with Father Teilhard and thus originated the invitation to come and help at Piltdown', wrote Woodward.

Of the outcome of that memorable encounter, which occurred in the spring or summer of 1909,[15] Teilhard de Chardin has himself written: 'Dans cette direction de la vieille Paléotologie Humaine ma première chance avait été (je l'ai dit plus haut) de me trouver mêlé, encore tout jeune, à la trouvaille en Angleterre, de l'*Eoanthropus dawsoni*.' The meeting in the quarry that Professor Teilhard recalls led not only to

Piltdown, but before that, through Dawson's good offices, to introductions to Professor Seward and Arthur Smith Woodward, who undertook the examination of the fossil plants and animals discovered by the two priests during the four years they were at Hastings, as a result of their assiduous search of the Fairlight cliffs. Besides numerous rare teeth of reptiles and fishes, the French priests succeeded in finding a second form of early mammal in the tooth of *Dipriodon valdensis*. This specimen, along with further teeth of *Plagiaulax* found by Dawson, were referred to Woodward at the end of 1909 and discussed in letters until Woodward's paper was ready in 1911. On 22 March 1911 Woodward gave his communication to the Geological Society. Dawson paid tribute in the discussion following the paper to the 'patient and skilled assistance' rendered to him by the two priests since 1909.

In November 1911 Dawson submitted to Professor A. C. Seward 'a small collection of plants obtained by him with the assistance of Father Teilhard de Chardin and Father Pelletier from the Wealden Beds of Sussex, for the most part in the neighbourhood of Fairlight'. Professor Seward's report to the Geological Society in November 1913 was given only after the two priests had finally left for France, and in their absence Dawson thanked the lecturer for undertaking the work, and once more paid tribute to the work of the priests and their generosity in donating the material to the British Museum.

On that very first occasion of the meeting in the quarry, Dawson revealed, with great enthusiasm, the exciting news of his find at Piltdown, and, as Professor Teilhard recalls, spoke of the several *pieces* he had already obtained, but Teilhard was not shown these first cranial pieces.

From the beginning of Woodward's interest in Piltdown, Teilhard took some part in the work. He was with Dawson and Woodward when the latter went down for the first time in June 1912 to view the gravels and to start excavations. That first day (or possibly the next) Teilhard alighted on a piece of 'Stegodon' tooth and the flint tool (No. 606). On one of those days, though it may have been later that month, Teilhard was present when Dawson found a piece of the right parietal bone. On three days in the summer of 1913 he worked with Dawson and Woodward. Returning to Hastings after a visit to Paris, he was invited by Dawson to stay at Lewes and to go on to Piltdown to help, and the next day, 30 August 1913, as we know, he discovered the canine tooth. In October of that year he was back in France. He did not return to England for many years. With the outbreak of the World War I he served as a padre with the French forces and from time to time he wrote to Dawson. After Teilhard's departure there were still further discoveries: the bone implement of 1914, and the second Piltdown man at Sheffield Park. In Paris Teilhard was associated with Marcellin Boule, the leading French physical anthropologist; he obtained his doctorate in 1920. From 1922 onwards he was for some years Professor of Geology in the Catholic University of Paris. He has made many notable contributions to archaeology and palaeontology, his speciality being fossil primates, and is well known also for his work in China, where he began work in 1924. In the Institut de Paléontologie Humaine he occupies the Chair of Geology as applied to the origins of man. He has not set out his views on *Eoanthropus* at any length,[16] but his assessment of Piltdown man, made in a letter[17] to Dr. Oakley, is that anatomically it constituted a 'kind of monster' and 'from a palaeontological point of view, it was equally shocking that a "Dawn Man" could occur in England'.

---

1. Weiner, J. S., Oakley, K. P., and Le Gros Clark, W. E., 'The Solution of the Piltdown Problem', *Bull. Brit. Mus. (Nat. Hist.) Geol.*, **2**, No. 3, pp. 139–46.

2. Dated to the Cretaceous (reptile) period, over 150 million years ago.

3. Now in the American Museum of Natural History.

4. Dawson, C., 1913, *Hastings and East Sussex Naturalist*, **2**, p. 76.

5. Woodward, A. S., 1916, *Geol. Magazine*, **3**, pp. 477–9.

6. Keith, A., 1950, op. cit., pp. 328, 654.

7. Hooton, E.A., 1946, *Up from the Ape*, p. 306, New York, Macmillan.

8. Straker, E., 1931, *Wealden Iron*, London, Bell.

9. Dawson, C., 1909, *History of Hastings Castle*, **2** vols., London, Constable.

10. Keith, A., 1950, op. cit., p. 324.

11. He had earlier proposed that Edgar Willett should drive them over: 'I have not told Willett anything about the situation of the gravel.'

12. Letters to Woodward, 20 April and 12 May 1912.

13. 1913, op. cit., p. 76.

14. The digging soon attracted local notice, arousing also the curiosity of the Uckfield police, Woodward wrote.

15. Dawson, C., in Discussion on Woodward's paper, *Proc. Geol. Soc.*, 22 March 1911. See also Woodward, A. S., 1911, 'Mammalian Teeth from the Weald of Hastings', *Quart. Journ. Geol. Soc. Lond.*, **67**, pp. 279–81.

16. See *Trans. New York Acad. of Sci.*, 1952, **14**, pp. 208–10.

17. 28 November 1953.

8

Some Others

¤ ¤ ¤

Teilhard de Chardin was by no means the first helper in the search. Very probably the first person to hear of Dawson's original fossil find, the piece handed to him by the labourer, was his friend of many years' standing, Mr. Sam Woodhead, a schoolmaster at Uckfield, who combined his teaching duties with the post of Public Analyst. Woodhead had carried out the analysis of the natural gas reported by Dawson to the Geological Society in 1898. He shared the first excitement of the finds at Piltdown, and went back to Barkham Manor with Dawson a few days after the first find to look for more fragments, but, as Dawson has told us,[1] their search was fruitless. Woodhead maintained his connection with the investigation, and it was he who carried out a chemical analysis of the skull[2] at some time before 1912. He remained at Uckfield till 1916, the year of Dawson's death. He was a man of considerable attainments, becoming a Doctor of Science and a Fellow of the Institute of Chemistry. He became Public Analyst for Brighton and Hove, and in 1916 went to live at Barcombe, the scene of Dawson's third discovery of human remains. He was among those who attended Dawson's funeral in Lewes in 1916. Woodhead often spoke of his early connections with the famous event to his wife and son[3] and to others, such as Mr. Essex, another teacher, at Uckfield. Mr. A. J. Smith of Leamington remembers in a conversation in about 1925 that, in telling of the event, Woodhead chuckled over his 'truancy' from school that day when he helped Mr. Dawson in the pit, as he did on subsequent occasions. These visits in 1908 are well remembered by Mrs. Sam Woodhead.

During the years from 1908 to 1911 Dawson showed one or more of the thick pieces of cranium to others among his friends and colleagues. Mr. Ernest Victor Clark[4] was given the privilege of a private view of the fragments when he and his wife were dining with the Dawsons in Lewes, at some time in the autumn of 1911 or early in 1912. He was taken down to the cellar and remembers that there were 'several, probably more than two, pieces of skull bone'. The Clarks understood that Smith Woodward was soon to take an active interest in the material as it promised to be of rare importance. Mr. H. J. Sargent, now Curator of the Bexhill Museum, recalls a chance meeting with Charles Dawson in Hastings in 1911 and seeing a piece of thick, brownish bone unwrapped from a piece of newspaper. Dawson said he was going to take it to the British Museum. In Hastings Dawson had a number of close friends and, after joining the local Natural History Society in 1907, he used to spend many Saturday afternoons in the Museum. He was on the Committee of the Museum, where he assisted with the organizing of loan exhibitions and where he deposited various antiquities on loan. A particular friend (and a frequent visitor to his home at Lewes) was the Curator, Mr. W. R. Butterfield. To him also Dawson showed the interesting object from Piltdown.[5] Butterfield claimed afterwards that it was he who advised Dawson to let the British Museum examine the remains. Also on the Museum committee and still very active at that time was that remarkable amateur geologist of St. Leonards-on-Sea, the jeweller Lewis Abbott.

Abbott's interest in this new discovery was intense. It expresses for us, vividly, the enthusiastic and also immediate acceptance which the Piltdown assemblage could evoke from many palaeontologists in those years. The Pliocene geology of the Weald and of Sussex had been Abbott's special study for many years. He was one of the Ightham circle which, in the 1890s, surrounded old Benjamin Harrison, the grocer and archaeologist of Ightham.[6] The curious formation of the Weald, the geology of the Downs, and the burning question of the human workmanship of eoliths were the common enthusiasms of the circle. Harrison's eoliths, like Reid Moir's Red Crag tools, persuaded most of the Ightham circle that Pliocene man (that is, man before the Ice Age) had existed and would be found. Abbott always maintained that the Pliocene formation would be recognized in the Weald. He claimed to have set Dawson to look out for this early gravel formation. He claimed, indeed, to have pointed the way to all the major discoveries in the south-east corner of Britain.

Abbott was a Londoner who as a jeweller's apprentice became infected with the Huxleyan vision of evolutionary biology. A self-taught man, an amateur like practically everyone in the Ightham circle, he earned his living both as a jeweller and as lecturer on gems in the newly-opened Regent Street Polytechnic. In the early 1890s he made a reputation with his Pleistocene finds[7] in the fissures near Ightham. These Shode fissures yielded over 100 Pleistocene species; previously only thirty-seven of such vertebrates had been known. For this he was given part of the Lyell Award of the Geological Society. He lived for a time at Sevenoaks to be near his palaeontological sites, but soon gave up his London post with Bensons. In the 1890s he took a jeweller's shop at 8 Grand Parade, Hastings, and from there conducted his multifarious geological and pre-historical pursuits, to the complete detriment of his business. He was the finder of the Hastings Midden Heaps, and pos-sessed a great collection of flints, animal bones, and some human and ape remains. A little, dark, black-bearded man, he was regarded almost as the oracle on everything that pertained to the geology of south-east England. His pronouncements were oracular indeed, and always of new discoveries, 'new races', and of 'new things in flints' which 'the world will have to rediscover them all', if they did not listen to him, as he wrote to his executor, Edward Yates, the antiquarian, of Hampton.

Abbott's business difficulties were well-known and his friends had to come to his aid more than once. Jewellery and palaeontology were inconsequently commingled at 8 Grand Parade. Abbott was a highly skilled watch- and clock-maker and worker in precious jewels. He offered for sale gems of any and every sort, including his own special 'proxy diamonds, guaranteed to natural gems the closest yet produced'.

He advertised 'New Lantern and Stereoscopic Slides' illus-trating prehistoric anthropology in every branch, 'Plateauliths, Palaeoliths, Mezzoliths, Neoliths', and all the unique speci-mens of Pleistocene vertebrates and the relics from the Hastings Kitchen Midden discovered by himself. At 8 Grand Parade in the 'Museum of Gems' there were on view examples of 'every gem known to have been used at any time for jewel-lery—matchless opals, phenomenal clear topazes, beautiful chrysoberyls, ideal cat's eyes, star rubies, gems of the Ocean, Diamonds from other worlds. . . .'

The jeweller's, now long gone, must have been submerged in a combined museum and old curiosity-shop stacked with flints and eoliths in great numbers, palaeontological specimens, animal and human remains, the many things obtained from the Fairlight Middens, the Sevenoaks barrow (which from Professor Graham Clark's scrutiny[8] must rank as one of the most confused excavations ever carried out), the Pleistocene mammalia from the foundations dug for the Admiralty Arch; there were Roman bronze statuettes, Roman pottery and tiles, Saxon spears, knives, fibulae and vases, medieval jars—as we may judge from the curious collection which Lewis Abbott put on show to illustrate 'Prehistoric Races' at the Hastings Museum Exhibition in 1909.[9] At this exhibition Dawson showed some iron work and wrote explanatory notes on the Sussex iron industry.

One of Abbott's exhibits, Item 6, calls for special mention:

> *Item* 6. *Human Teeth.* Some worn down by gritty foods, some jaws show abnormal dentition. In one case the last molar is more than twice the size of the first—an essentially pre-human character; another case shows shortening of the jaw at the expense of the number of teeth. A normal jaw is shown for comparison.

This molar 'twice the size of the first' and its lodgement in the jaw would take any anatomists' immediate attention, for, I venture to say, it is unheard of in a normal adult jaw—unless Abbott was quite unable to distinguish between permanent and milk molars in a child's jaw!

Abbott's Pliocene and Pleistocene collections are to be found in many museums in the country, but of his stock of human bones nothing has so far been traced.

In his day the views of Lewis Abbott carried weight. Over

the matter of the Piltdown eoliths,[10] Dawson at first expressed himself with some circumspection, making it clear that this was a vexed subject, but one on which Lewis Abbott would in due course help to throw light, as he had considerable collections and experience of eoliths. Abbott was a firm believer in the human manufacture of eoliths. Abbott's reputation on matters palaeontological and archaeological stood so high that when Dawson and Woodward made their finds of flint implements, it was Abbott whom Dawson sought out to consult about their authenticity and joyfully reported the verdict in a letter of June 1912 to Woodward: 'Abbott is in no doubt. They are man, and man all over.' On the geology of Piltdown, Abbott was equally positive, and Keith, as we have seen, quoted Abbott's verdict as to the Pliocene age of the lowest stratum in the gravels in support of his own preference for a date earlier than Dawson's.

On this matter of the gravels, Abbott took the opportunity of pressing his views in a newspaper article[11] in the February immediately after the first Piltdown meeting of 18 December 1912, and again in an open-air discourse to the Geologists' Association when they picnicked at the famous spot in July 1913, an occasion when Keith noted in his diary: 'Abbott was everywhere in evidence.' Abbott went further; he claimed a place in the great discovery. He insisted in the article of 1 February that it was he who had brought to the notice of his 'colleague on the Museum Committee' the gravel spreads of the Weald as likely places to look for the remains of fossil man, that when he had aroused Dawson's interest, thorough investigation soon produced the 'spoil which he brought to me from time to time'. Abbott, like Woodhead, Sargent, Butterfield, and Ernest Clark, had also been shown the first Piltdown fragment soon after its discovery, and he was impressed, he writes, with its 'superlative importance'.

In this article of 1 February 1913 in the *Hastings Observer*, Abbott dogmatizes at some length on the anatomy of *Eoanthropus*, declares the skull to be 'a mixture of the human, the gorilloid and chimpanzoid' and expounds the essentially human features of the 'chimpanzoid' jaw. Though many of the comparisons he makes are questionable, it is clear that he was quite familiar at that early date with the character of the Piltdown jaw. Some of what he wrote he could no doubt have learnt from attendance at the December meeting,[12] and also from the uncorrected proof of Dawson and Woodward's abstract distributed on 18 December to the Fellows of the Geological Society. Dawson and Woodward's detailed paper did not appear till March in the *Quarterly Journal of the Geological Society*, and Dawson's own version was printed in the *Hastings Naturalist* on 25 March. The only reconstruction of *E. dawsoni* available at that time was the one exhibited at the meeting. Other casts were not reproduced till April or May of 1913. Thus neither of these 'official' accounts was available to Abbott for his article of 1 February, nor could he have obtained from attendance at the meeting some of the information he gives on 1 February. He must therefore have had some opportunity of handling the actual jaw or its cast in 1912. This is not really so surprising, as we know that he was called into consultation over the flints in June 1912 when Dawson went down to see him. His version of the circumstances of the discovery, given on 1 February, differs materially from the account he would have heard on 18 December and is in important essentials similar to Dawson's account in the *Hastings Naturalist* of March 1913. There is thus much to suggest that Abbott was in close touch with Dawson, and Mr. Yates, Abbott's executor, confirms this was so.

So important did Abbott[13] feel his contribution to be that he wrote personally to Woodward a few weeks before the December meeting of 1912 to point out with some vehemence that Dawson would not have made the discovery but for his inspiration and instruction.

Abbott always maintained a keen interest in the affair. As already recorded he took an active part in the excursion of the Geologists' Association to Barkham Manor on 12 July 1913, organized by Dawson. He regarded it as an historic occasion, writing to Yates to find out what notice had been taken of it by the 'national picture-papers' and in the local press (Yates had a relative at Uckfield). He urged Yates to ensure a wide distribution of the photo (Pl. 8) which Yates had taken. 'I expect Keith, Corner and others next Sunday to hear more about what I told them at Piltdown', but of this sequel to the excursion no information is available.

In 1914 we do not hear of him in connection with Piltdown, but in 1915 he took strong exception to the attack on the 'eolith' theory made by Dawson at the Royal Anthropological Institute in February[14] and wrote 'abusive letters' to Dawson.[15]

It is likely but not certain that he knew of the finding of the second lot of Piltdown fragments in 1915. In a letter he wrote from a sick-bed to Yates in February 1917 he exclaims: 'Oh! did you see that those other fragments of a second Piltdown skull were described last Wednesday by Smith Woodward at the Geological?'

In later years he still alluded to his own part. He told a number of visitors on one occasion,[16] almost certainly in June

1929, that Dawson at first thought the Piltdown skull was of the nature of an iron-stained concretion, but that he, Abbott, had persuaded him that it was a genuine fossil bone. To Mr. Edmunds,[17] mapping the Lewes area in 1924, he imparted the information that he had worked with Mr. Dawson on the Piltdown skull, that it had been in his possession six months before Woodward saw it, and that they had soaked it in bichromate to harden it.

In 1915 the war brought a huge camp of Canadians[18] and, with their free-spending, things looked up for Abbott, but it did not last. From 1920 onwards he was constantly beset with money troubles, and he had several times to appeal to Yates. A subscription in 1921 organized through Butterfield helped him, and another was raised ten years later by some Fellows of the Royal Society. Keith went down to present the money, but found Abbott *in extremis*. Abbott died in 1933 aged eighty, and Keith wrote an obituary for *The Times*.

Lewis Abbott is remembered still for the fiery, bombastic, inspiring, and weird character he was. It is said that he would never divulge the source of many of his finds, would let no one handle his exhibits, and was possessive of his 'rights' over the Fairlight and neighbouring sites; today his stock has fallen— the kitchen middens do not at all conform to his exuberant description and they are no more than the medieval castle's refuse; the round barrow and mesolithic mixture at Sevenoaks defies understanding; the 'Red Crag' of Sussex existed only in his imagination—his section of 'boulder clay' just outside Hastings has proved to be nothing more than a layer of beach pebbles and part of an old road.

Abbott's whole attitude to the Piltdown discoveries testifies eloquently to the accomplishment of the Piltdown hoaxer in

bringing to realization a set of ideas and theories already entertained, in lesser or greater degree, by others. We know that in
old Benjamin Harrison's circle there was always talk of the
possibility of finding a late Pliocene deposit in the Weald or on
the Kentish Plateau. We know of searchings which were made
in the last decade of the century of the water partings in the
Weald near Ightham and the ridges between the combes of the
North Downs for traces of gravels deposited by pre-existing
streams. Old Ben and his friends (of whom F. J. Bennett was a
particularly keen searcher) had many a walk to likely spots, but
they were never successful.

Once excavations got under way at Barkham Manor, there
were some visitors, even in 1912 before the public announcement. Keenly interested spectators were Mr. Kenward, tenant
of Barkham Manor, who had granted permission for the work,
and his daughter, Mabel Kenward. Before she went away on
war work in 1914, Miss Kenward and her friends kept in touch
with the exciting doings at the pit. Besides the principal workers and Hargreaves the labourer, we know now of no other
excavator.[19] A not infrequent visitor appears to have been Mr.
Edgar Willett, a friend of Dawson's who gave much help in the
search for the further extension of the gravels.[20] Abbott was
there at various times.[21] One visitor who came back two or
three times in 1912 was none other than Sir Arthur Conan
Doyle. Perhaps the Piltdown discoveries provided him with
some inspiration, for at this time he was busily writing his *The
Lost World.* Dawson found his visits most gratifying and wrote
to Woodward in 1912: 'Conan Doyle has written and seems
excited about the skull. He has kindly offered to drive me in
his motor anywhere.' Of those who soon became publicly
associated with discussion on Piltdown man[22] none is known

to have visited Barkham Manor in 1912 before the first announcement.

Casual and serious visitors were there in great number after the discovery became public; many geologists, anatomists, palaeontologists, and archaeologists went down for a day's visit to inspect the site. Of these many visitors we may mention Captain Guy St. Barbe, who remembers how surprised he was to find the site so extremely shallow and restricted, and Major A. G. Wade of Farnham, who felt sceptical of the geology and shared his doubts on the matter with his friend, Mr. Reginald Smith, archaeologist at the British Museum.

Who the labourers were who in 1908 found and broke what they took to be the 'coconut' we do not know. Possibly one of the two men was of the Stephenson family, who have always worked in the quarries in and about Piltdown. Tom Paget (who has been mistakenly supposed to be one of the original labourers) was working on the farm in those days. He was helping to lay gravel on the road very near the excavators' pit in 1912 and remembers the excitement when Dawson found another piece of the cranium. From Mr. Paget's description, which is quite clear, this piece was without doubt the right temporal which we know in fact to have been found at that time.

The scene at Piltdown cannot be followed in any great detail as we do not possess the necessary information of the way the work was conducted in the years 1912 to 1916. What is quite certain is that the famous place was readily accessible to visitors and others. The introduction of forged and extraneous material into the pit would have been easy enough. For years before 1912 it was known to not a few as the place where Dawson had stumbled across 'the oldest known flint gravel in the Weald', and from which he had recovered, long before 1912, a 'part of a human skull to rival *Homo Heidelbergensis* in solidity'.

---

1. 1913, op. cit., p. 76.

2. Dawson and Woodward, 1913.

3. Dr. L. S. F. Woodhead, M.B.E.

4. Mr. E. V. Clark died aged eighty-six in February 1954.

5. Letter from the Rev. S. G. Brade-Birks, 19 February 1954.

6. The biography of this remarkable amateur has been written by his son, Sir Edward Harrison: *Harrison of Ightham*, 1928.

7. Abbott, W. J. Lewis, 1894, 'The Ossiferous Fissures in the Valley of the Shode near Ightham, Kent', *Quart. Journ. of Geol. Soc.*, **50**, pp. 171–87.

8. Clark, J. G. D., 1932, *The Mesolithic Age in Britain*, 223 pp., Cambridge.

9. *Catalogue of an Exhibition of Local Antiquities*, Corporation Museum, Hastings, 1909.

10. Dawson and Woodward, 1913, op. cit., p. 122.

11. 'Prehistoric Man. The Newly Discovered Link in His Evolution', 1913, *Hastings and St. Leonards Observer*, 1 February 1913.

12. There is no certainty about his attendance. The surviving register does not bear his signature for that meeting, but there may have been another register.

13. Letters dated 24 November 1912 and 15 December 1912.

14. 'Sussex Ouse Valley Cultures', read at Royal Anthropological Institute 23 February 1915; not published.

15. Letter of Dawson to Woodward, 9 March 1915.

16. Information from J. Jackson in letter from Dr. W. D. Lang to Mr W. N. Edwards, 30 November 1953.

17. Personal information.

18. Including Grey Owl (Archie Belaney), who was born at Hastings.

19. A Miss (or Mrs.) Simpson is said to have helped, but no verification has been obtained.

20. Dawson and Woodward, 1913, op. cit., p. 151.

21. According to Captain St. Barbe and Mr. R. Essex, but Professor Teilhard de Chardin is doubtful.

22. Apart from Woodward and Dawson, this group would include Dawkins, Elliot Smith, Lankester, Keith, Kennard, Reginald Smith, and Pycraft.

9

Lineaments of the Forger

¤ ¤ ¤

If we hope to identify the perpetrator of the Piltdown fraud, or to get near doing so, we may begin with some quite obvious questions. This may help us to sketch out what manner of man we are dealing with and perhaps give us a few clues.

How could the faked jaw have been obtained? We can at once dispose of the notion that any difficulty would have been experienced in procuring an ape's mandible, or in all probability several, for the forger may have used up a number before he was satisfied. Then, as now, they could be bought from, or through, a local taxidermist, or, if not, then easily enough from one of the famous London firms. One would almost certainly go to Gerrard's in Camden Town, who have been established in this business (in their Camden Town address) since 1860. Anyone who has ever worked in a museum would know of the firm. Mr. Gerrard tells me that unmatched jaws and other odd bones were probably easier to come by in the years before World War I than now. Before the blitz of the last war blotted out their premises, once a month one could attend a 'taxidermists'' auction in the rooms of Stevens in King Street, and a collector could hope to pick up an enormous variety of specimens and bones. Odd bits of the skeleton, such as teeth and mandibles, were cheap enough in those days. Ape and human jaws could be easily come by and many geologists had them. Lewis Abbott, for example, when he wrote of the Piltdown jaw in 1913, had access to skeletal material for his comparative study; he mentions some of the simian features of the fossil jaw as 'obtaining in a chimpanzee jaw now before me'.[1] The field of inquiry is therefore not particularly narrowed down by a consideration of this sort.

Flints like the 'worked', so-called, pre-Chellean flint implements at Piltdown could also have been obtained or even made without particular difficulty; but at once we are aware that we are dealing with a reasonably informed archaeologist. If the 'ancestral' Piltdown man was to be passed off in form and in date as very ancient indeed, something reminiscent of the Red Crag, his tools would naturally have to be of the right style for the time. A slightly later style (Chellean or Abbevillian) might have served, but would be risky, since the workmanship might be thought too advanced for the apish man of possible Pliocene date. Moreover, the style endured for so long in the geological record; there might be a risk of the hoped-for date being misunderstood. Making the tools pre-Chellean, and therefore much ruder and definitely early, was a good stroke. The flints, as we now believe, were really rejects off a Neolithic block, and they were stained to give an appropriate iron patina and to conform with the colouring of the stones of the gravel spread. We realize from this that the archaeologically minded perpetrator was well acquainted with the Barkham geological strata and must have been so early on.

Both these deductions, the professionalism (indeed virtuosity) of the culprit and his close acquaintance with the gravel formation are confirmed repeatedly from a consideration of the rest of the spoil. In addition to his understanding of archaeological matters he can be confidently credited with considerable palaeontological knowledge and experience. The unique bone implement has been fashioned from an elephant femur of a type early enough to be thought antecedent to the advent of Piltdown man, its maker. Woodward always thought of the elephant tool as the most convincing of the testimony to Piltdown's antiquity. The extraneous fossil fauna, too, have been chosen to provide an identifiable stratification of the 'Upper Pliocene' and early Pleistocene, showing levels both earlier and contemporaneous with the Dawn Man. Thus the culprit must have known what fossil elephant, mammoth, rhinoceros, and hippopotamus teeth to get and to know them for what they were. The rhino and mastodon are particularly well chosen, for they are authentic Reg Crag fossils from East Anglia. All this suggests an established collector, someone able to find, buy, or obtain by exchange quite a number of Pleistocene fossil specimens. But we must be careful not to exaggerate the extent of these transactions. Among the mammalian fauna there are only a very few of the really rare ones—at most three in all, and these have been broken so that a number of separate fragments of each specimen were found at different times. The 'cache' of 'planted' fauna, as we have seen, is a strange medley and the occurrence in it of Mediterranean forms—the 'North African' elephant and the 'Maltese' hippo—suggests their acquisition by the perpetrator from foreign collections.

The culprit, we may be sure, had the actual handling of the cranial fragments, in order to match the spurious jaw closely in colour. This implies that the fragments must have been in his hands at some time between 1908 (or even earlier) and June 1912, when the jaw was found; but we are not helped much by

this consideration, as we know that Dawson showed the specimens round to many acquaintances before they came to
Woodward. The guilty person continued, of course, to keep in
close touch with events subsequent to that first season of finds.
He must have known of the crucial importance which had
come to be attached to the finding of the missing eye-tooth. As
the tooth forms one of the mounting sequence of finds, he
must have been kept well acquainted with events at the pit
itself. He must have been able to arrange for the finding of a
number of objects in 1914 (including the bone implement)
and in 1915 at the second site at Sheffield Park. The neighbourhood there was explored on several occasions by the party
from Barkham[2] in the spring and autumn of 1914; Teilhard
was conducted to the place by Dawson in 1913;[3] these excursions and Dawson's explorations with Edgar Willett for the
further extension of the hitherto unmapped Ouse gravels
would have been common knowledge and so made it possible
for 'finds' to be left in these promising sites.

The perpetrator can be delineated as someone long aware
of the potentialities and characteristics of the gravels and in the
closest touch with every event in the history of the discoveries,
and someone both archaeologically and palaeontologically
well-versed. There is nothing particularly surprising that such a
man should think of the initial deception, for the inspiration
can well be understood as the inevitable outcome of the finding of the ancient Ouse gravel in the years of the Heidelberg
jaw, already well known by the end of 1907. Given a gravel
terrace older than that of Heidelberg, what more logical than
for the deceiver of *posit* a creature even more ape-like? The rest
would follow—for the Mauer jaw is ape-like in the chin
region, but human in its teeth (nearly all of which are

preserved), and the articular condyle is human. Any reader of the *Quarterly Journal of the Geological Society* would also know that a broken condyle was not unusual, for the Cheddar man of Upper Ice Age date described and pictured in 1904[4] had just such a breakage.

The remarkable array of specimens from the Piltdown sites makes the perpetrator appear, at first sight, a man of rather extraordinary talents. He seems to possess the abilities of an expert palaeontologist and geologist, as well as to be highly adept in chemistry, human anatomy, and dentistry; yet this would certainly be an uncritical and exaggerated assessment of his qualities. Without doubt the key to all his accomplishments lies in a solid palaeontological background or training. Proficiency in this goes far to explain all that was done at Piltdown. With his palaeontological knowledge, the perpetrator would realize (and, as we know, he would not have been alone in this) the potential significance of an apparently Lower Pleistocene or Upper Pliocene gravel deposit. His experience would tell him at once with what kind of animal fauna to stock this horizon. He would be aware of the relatively simple archaeological succession recognized in 1908, and therefore would know the likely tools to expect in his gravel deposit. The idea of a possible ape-like man at this period would be a completely familiar notion to any well versed palaeontologist. For such a man, well experienced in handling fossil mammalian material, little specialized knowledge of human anatomy would be required for putting into operation his great plan. In the Heidelberg jaw and in the few remains of Java man there was a clear and easily comprehended guide to the probable features of the missing ape-man. Subtleties of human anatomy and dental anatomy would not then (or even now) deter the experienced palaeontologist from handling human fossil material. We know, for example, that Smith Woodward, despite his life-long preoccupation with fossil fish and reptiles, did not hesitate at all to undertake an analysis of Piltdown man. His colleague, Pycraft, an ornithologist, embarked on a comparative anatomical study of the human and ape mandibles and the Piltdown specimen. We have seen even an amateur palaeontologist like Lewis Abbott expatiating on the essential human and ape features of the newly discovered *Eoanthropus*, and the reader will recall his uninhibited claims for the human specimens he exhibited at Hastings in 1909. For that matter, Dawson, as we know from his letters,[5] was quite ready to pronounce on the resemblances of Piltdown man to the Cheddar Gorge specimen or the Aylesford skull,[6] and to join issue with Keith[7] on the characters of the Piltdown jaw and teeth and to undertake an inquiry (in 1912) into peculiarities of the human vertebral column, stimulated by a specimen which had taken his attention in the Museum of the Royal College of Surgeons. To this day it is to the general palaeontologist that we owe much of our knowledge of (for example) the fossil ape bones of India or of the more recently discovered representatives of Java man.

The brilliant idea of abrading the ape teeth does not bespeak a very special knowledge of dental anatomy. For one or both of two reasons, it would arise almost inevitably from the initial idea of combining an ape jaw and human cranium. The condition of the teeth in the Heidelberg jaw (and the aim was to better this specimen or imitate that in the molars of Java man) would have provided sufficient inspiration to so careful and determined a worker as our man undoubtedly was. But in any case he could not possibly have left the molar teeth of an orang untreated, for the occlusal surface of the molar teeth are so different, not in subtle features of cusps, but in the most obvious way. The molars of the orang or chimpanzee present to greater or lesser degree a pattern of crenulations so very different from anything on the surface of human molars that they would certainly have to be removed, as removed they were. The remains of these crenulations are to be seen in the form of little pits and crevices on the Piltdown molars similar to those on the orang teeth abraded by us to simulate the originals. Nor is there a special mastery in the art expended on the canine tooth. This tooth, as we know, only turned up when it was clear that the experts had decided what it would look like. The canine found by Teilhard had been subjected to abrasion to reproduce a shape which had already been 'predicted' and modelled in the plaster-cast specimen exhibited by Smith Woodward for some six months before the finding of the real tooth. To this day it can be matched closely only by this plaster antecedent.

The way in which the Piltdown specimens were treated thus reflects quite clearly the influence of contemporary models furnished by other fossils. The specimens which came to light in the early stages have been treated to conceal as much as possible of those features which might have drawn doubt on the evolutionary importance of the cranium or the jaw. The frontal bone of Java or of Neanderthal man carries a well developed eyebrow ridge and this region was almost completely missing in the first Piltdown finds. When the experts in 1912 decided that *Eoanthropus* must have possessed a modern forehead and that it gained a special evolutionary significance from this very fact, the appropriate piece, with its 'angelic' forehead, as Dawson put it,[8] was forthcoming at Piltdown II.

In the mandible, apart from the abrading of the molars, the chin region has been broken off, again with the idea of making it difficult to diagnose its essentially ape-like characters. The hoaxer certainly succeeded in making it impossible for general agreement to be reached on the correct reconstructions of the jaw and cranium and so brought about much acrimonious argument amongst human anatomists. But he did not prevent Woodward from diagnosing the predominance of ape characters in the jaw and in the missing canine. If the perpetrator had really been a highly trained human anatomist, he would probably have disguised the forehead and chin regions rather more cleverly (though we may suppose that a shortage of material may have prevented him from doing a more thorough job).

Finally, as to his attainments in chemistry. Here again his activities do not necessarily reflect anything more than his abilities as a good geologist and palaeontologist. It is not in the least remarkable that he should be aware that fossil specimens are often iron-stained and that the gravel should be iron-bearing. It was, therefore, only a matter of his using one or two iron salts for his staining experiments, and incorporating dichromate, an obvious oxidizing agent.

We require in our perpetrator no special technical qualities other than those of a sound, well-versed geologist and palaeontologist; but these were certainly combined with high personal distinctions of imagination, perseverance, and boldness in pursuing a clearly conceived and firmly held objective. Behind it all we sense, therefore, a strong and impelling motive.

The planning of a sequence of events of this degree of elaborateness, of a watchfulness of the reactions of the scientific community from the days of the first meeting onwards, must betoken a motive more driving than a mere hoax or prank. A hoax could have been exploded, called off and its object attained in ample time between June 1912, when the British Museum became involved, and December 1912, when the affair became public. If the hoaxer felt he dared not make his confession, why should it be carried on any further and for three more years with even more startlingly novel finds, all carefully timed? Still, it cannot be ruled out, for the joker might have had the pseudo-scientific aim of discovering how far the palaeontologists could be taken in by building up a complete horizon for them.

Rather stronger in its appeal than a mere jape of the Horace Cole school is that of the working off of a grudge against one of the principals in the discovery with a view to ultimate exposure. This could account for the elaboration and persistence of the deception; moreover, Dawson, for one, as we shall see, had enemies or anyway ill-wishers amongst his colleagues. This motive too seems improbable. The perpetrator, himself a person of some professional standing, would have considerable difficulty in 'exposing' the plot without involving himself. In any case, we know of no attempt at a dénouement, even though there were those who from the earliest days regarded the events at Piltdown as in all likelihood fraudulent.

There are, of course, other and more obvious motives. There could have been a mad desire to assist the doctrine of human evolution by furnishing the 'requisite' 'missing link'. With the general truth of biological evolution so well attested, with signs in abundance of tool-making man at the earliest Pleistocene times, if not earlier, the gravel of that very date at Piltdown might have offered an irresistible attraction to some fanatical biologist to make good what Nature had created but omitted to preserve.

Finally, there is the simple impulse of personal ambition, one not incompatible with the preceding motive. The world-wide fame of the pioneer discoverer of 'missing links' would be vivid to the generation of the Piltdown explorers. Dubois's epic discovery was a bare twenty years old, Otto Schoetensack's Heidelberg man very recent. The impulse of ambition could have been alloyed with a resentment at insufficient recognition imagined or real of the perpetrator's achievement in palaeontology.

Dawson's death coming so soon contributed greatly to the perpetuation of the secret history of the affair. If, as in Kipling's story,[9] revenge was the aim, or blackmail, or a grand spoof, or anti-evolutionism, the passing of Dawson would remove the incentive and a dénouement would now become inconvenient or dangerous to the informant. Motive, at this stage, does not take us far. It indicates only that among the Piltdown group we shall not be altogether surprised if the perpetrator turns out the victim of an associate, or even a blackmailer, an unhinged evolutionist, or a man of overween-ing ambition. It would be idle to cast those parts even though it is perhaps just possible to do so. What is reasonably certain is that the culprit, in view of his manifest characteristics, can hardly fail to be among those whose names we know. It seems in the highest degree improbable that someone, with the qual-ities and knowledge we have inferred, so closely connected with the affair, over so long a period, should somehow escape all mention in the contemporary accounts or in our recent inquiries.

With this general consideration in mind we may now

reconsider the events in Sussex in greater detail, using the accounts given by Woodward and Dawson, the latter's correspondence, and information supplied by Professor Teilhard de Chardin and other sources.

---

1. Some 'monkey-bones' or 'skulls' are said to have been found in Dawson's house in Lewes during repairs in 1931–2, but, despite careful inquiry, no confirmation of this report has been forthcoming.

2. Woodward, A S., 1917, op. cit.

3. Personal communication.

4. Davies, H. N., 1904, 'The Discovery of Human Remains under the Stalagmite Floor of Gough's Cave, Cheddar', *Quart. Journ. Geol. Soc. Lond.*, **60**, p. 335.

5. Letter to Woodward, 30 June 1912.

6. The whereabouts of this skull cannot be traced.

7. Letters to Woodward of 31 October and 26 November 1913.

8. Letter to Woodward, 9 June 1915.

9. 'Dayspring Mishandled' in *Limits and Renewals*, 1932, Macmillan (the story was first published in 1928).

10

Events Reconsidered

¤ ¤ ¤

It is unfortunately not possible to follow in any detail every stage of Smith Woodward's activities at Piltdown. No diaries or note-books exist of the work done, there is nowhere a complete record of the various finds as they were made. Woodward kept copies of very few of his own letters and we have only the letters written to him and now preserved at the British Museum. When the American palaeontologist Osborn came over in 1920, Woodward dictated some notes which help to allocate the various discoveries. Apart from these notes and the one-sided record of the correspondence, there are only the reports in the scientific literature and popular lectures on Piltdown as primary sources.

Woodward does not appear in general to have been a secretive man, but over the Piltdown material he went to some lengths to keep the whole affair as quiet as possible until near the time of the public meeting in December 1912. He did not consult any of his colleagues in the Museum about the finds or about the interpretation he was to place on them. Mr. Hinton says that to his colleagues at South Kensington Woodward's diagnosis of *E. dawsoni* came as a surprise mingled with some dismay, for there was much scepticism of the new form amongst his museum colleagues, including Oldfield Thomas and Hinton himself. They would have advised caution, he says. Keith knew nothing of the events in Sussex until rumours reached him in November. He wrote asking for a view of the exciting material, but on his visit on 2 December to the Museum he was received rather coldly and allowed a short twenty minutes. But, judging from Dawson's letters in 1912, it seems fair to say that Woodward was merely seeking to avoid a premature disclosure, for he had decided early on that Piltdown would indeed prove a sensational event. Woodward did not want any of Dawson's 'lay' friends to come along on his first visit to the gravel when he had yet to make up his mind about the real importance of Dawson's find and of the necessity for systematic excavation. He was willing to take Teilhard along, of whom Dawson had written that he was 'quite safe'. He knew, of course, of Teilhard's geological enthusiasm, as he had himself examined the young student's finds from Fairlight in 1910 and reported on *Dipriodon* in March 1911.

Woodward gives few dates, and nowhere mentions that the series of Piltdown discoveries began as far back as 1908. In fact, in the *British Museum Guide* of 1915 and again in his *Earliest Englishman* he speaks quite loosely of 1912 as the date of the discovery, when, of course, before then some of the material had already been unearthed. Most surprising are his contradictory accounts of the finding of the canine. When he announced the exciting news of the discovery of the canine at an evening discourse at the British Association meeting at Birmingham on 16 September 1913 and later in a Friday evening discourse at the Royal Institution in December 1913, Woodward does not say that he was present on that occasion. He wrote: 'Fortunately, Mr. Dawson has continued his digging at Piltdown during the past summer and on 30 August Father

Teilhard, who was working with him, picked up the canine tooth.' Yet in the *British Museum Guide* of 1915 and in the first chapter of *The Earliest Englishman*, which Lady Smith Woodward informs me was written in the Piltdown days (though only taken up again in 1944), Woodward makes it plain that he was there, with Dawson, when Teilhard found the canine: 'In the following season—1913—we continued to work without much success until Saturday 30 August, when we were accompanied by Father Teilhard.' This confused recollection or careless recording on the part of Woodward is rather unfortunate when we try to elucidate the circumstances in which the canine might have been planted for Teilhard to find it on 30 August. But Professor Teilhard tells us that Woodward certainly was there, and (after congratulating Teilhard on the keenness of his observation) that he put the tooth in his pocket. At this time (and again in 1916) the canine seems to have puzzled Smith Woodward somewhat, even though it conformed closely to his expectations. The two letters from Dawson in October and November 1913 contain data which are clearly intended to reassure Woodward on some of the peculiar features of the tooth. Dawson draws his attention to the occurrence of 'a female gorilla tooth at Keith's' (the Royal College of Surgeons) 'amazingly like ours but larger . . .'. He gives a detailed description and encloses some good sketches.

> When in position in the jaw [the canine] does not show the worn surface . . . the portion of the tooth below the worn part is polished but little worn at all. The wear goes right down on to the upper end of the pulp cavity rather more than ours. . . . The wear has given the tooth a lancet shape like ours.

And on 26 November he sends Woodward 'squeezes' of the

female gorilla for Barlow to make casts and photograph. But nothing of this gorilla counterpart to the Piltdown canine apparently carried enough weight to figure in Woodward's description of the find. Keith at this time took an extremely critical view of the canine; he could not believe it belonged to the jaw, as the wear seemed to him totally excessive in a jaw in which he judged the third molar not to be fully erupted. Dawson warmly contested Keith's arguments, as he recounted to Woodward in some detail. The fact that Keith did not pursue his contentions was largely due to the opposition he encountered from the dental expert, Dr. Underwood; it probably owed something as well to only one of Dawson's arguments—namely, the fact that 'we have two molar teeth worn quite as much as the canine'.

But in February 1916 Woodward raised in private with Dawson the very point on which W. Courtney Lyne had (respectfully) challenged the eminent protagonists of *E. dawsoni* at the Royal Society of Medicine some weeks before. On that occasion the dental practitioner had obtained no real satisfaction in reply to his reiterated challenge:

> I would have liked to have seen thrown on the screen tonight a picture showing anything like the amount of wear and tear (in a tooth) such as the Piltdown tooth displays, and with a similar cavity. You will find when the wear is very heavy that the pulp cavity is practically obliterated.

Woodward poured scorn (in general terms only) on Lyne's assertion. It was, of course, Underwood's views which bolstered Woodward on this point and Lyne's lone foray failed to shake *Eoanthropus dawsoni* and was forgotten. Yet on 16 February Dawson writes to Woodward:

> The pulp cavity is certainly large. It does not seem to have occurred to anyone that as one end is open the walls of the cavity may have been the subject of post-mortem decay and that bacteria may have cleared away the comparatively soft walls during a prolonged soakage in water and sand. I think I have noticed this in fossil teeth and broken bones.

The great significance of this correspondence lies in this—that for all his convinced manner and crushing method with those holding opposing views (as with Keith no less than with Lyne) Woodward was not so irrevocably committed to *Eoanthropus dawsoni* as never to raise some queries, at least in discussion with his colleague.

Woodward does not tell us in so many words exactly what it was that Dawson brought him on 24 May 1912. In the posthumously published book[1] of 1948 we are told that four of the nine cranial fragments were found after May 1912; we infer that Dawson had brought along five cranial fragments. But neither Dawson's nor Woodward's published articles contain sufficient information to provide really detailed confirmation on this point. Some independent support can, however, be adduced. The obituary notices on Dawson in 1916 speak of a *third* fragment as found in the autumn of 1911, and Mr. Clark of Lewes at about that time, as we know, was shown in the cellar 'several, more than two fragments'. Teilhard, too, is under the impression that Dawson had several pieces before going to the British Museum, as he learnt on his first meeting with him.

Woodward mentions, again only in 1948, that in addition to the human fragments, Dawson brought two hippopotamus teeth, one of 'stegodon', and some flints. There is complete confirmation for the hippopotamus pre-molar, since Dawson had already written about it as early as March 1911. As to the 'stegodon', one can be reasonably sure from the published accounts and from the Museum register that one piece was brought along. It is recorded that three more pieces were unearthed after Woodward's participation. The British Museum does in fact possess three large pieces and a fourth rather smaller splinter. As to the flints, Woodward does not say in *The Earliest Englishman* exactly which of these came into his hands in May, but he refers to them as flint tools, so that they were obviously not eoliths, which he nowhere refers to as tools. The implication is that Dawson had already had at least one of these worked tools. It must be emphasized that despite our fragmentary information we have therefore substantial independent confirmation of what Woodward said he had obtained from Dawson before June 1912. We can be fairly sure of at least three cranial pieces, some of the tools, and the 'stegodon' tooth, and can be quite definite about one of the hippopotamus teeth.

Nowhere does Woodward say when it was that he first became aware that Dawson had stained the five fragments brought to the British Museum on 24 May 1912. The staining is mentioned in print for the first time only in 1935, nearly twenty years after Dawson's death, by Woodward and Hopwood, neither of whom attach much importance to it, as the stain had affected the colour to a minor degree. A difference in colour is very noticeable even now; the temporal bone is quite yellow, but the variation could be taken as natural. Sir Arthur Keith cannot say when it was that he first heard of the staining, though he had known about it for a considerable time. It is not mentioned in the 1915 or 1925 editions of the *Antiquity of Man*, but he believes he heard of the staining from

Dawson himself. If Woodward knew that the bones had been kept in bichromate solution and in Abbott's possession in Hastings some six months before he himself came into the affair, it certainly was not regarded as of sufficient importance for him, Dawson, or Abbott (in 1913) ever to mention the circumstances.

It cannot be gainsaid that there is a disconcerting vagueness in Woodward's accounts on some of the circumstances of the affair, the sequence of finds, the dates of discovery, and the details of the chromate-staining. But unsatisfactory as his account appears to us now, it is surpassed by the obscurities and inaccuracies of Dawson's description of events.

Like Woodward, Dawson does not expressly mention 1908 or any date as the date of the first cranial find, even when, as before the Geological Society, he is purporting to give the 'official' history of his discoveries. This date we infer from several of the obituary notices of 1916 (Keith records it is 1925, but confuses it with the date of finding the gravels) and it is confirmed by Mrs. Sam Woodhead and her son, Dr. L. Woodhead, as we have noticed already. Dawson, speaking on 18 December 1912, merely says that the first fragment was given to him 'several years before' he recovered the second piece in the autumn of 1911. The 1908 date seems the most probable date, but it is not absolutely certain.

Obscurity also envelopes the date when, attending as Steward of Barkham Manor, he originally noticed the all-important gravel terrace itself. In his 1912 paper Dawson says that the event occurred 'several years ago' at some unspecified time 'shortly' before the workman handed him the first piece of parietal. The account in the *Hastings Naturalist*, which was certainly prepared at much the same time (it appeared on 25

March, in the same month as that in the *Geological Journal*), gives a different impression. Instead of a date 'several years' before 1912, here he says quite definitely (and Woodward repeats this in *The Earliest Englishman*) that the peculiar gravel came to his attention at Barkham Manor shortly 'before the end of last century' some thirteen years before. This is just what Abbott had explicitly stated in his article of 1 February 1913, which came out long before Dawson's. On that same day (in the geological paper he says 'shortly aftewards'), Dawson visited the gravel pit, and asked the two farm hands at work there to keep a sharp look-out for fossils.

Now, the occasion of Dawson's visit to Barkham Manor was the periodic Court Baron over which he, as Steward of the Manor, presided. These were held approximately at four-year intervals.[2] The records of the Barkham Manor Estates are available.[3] As one took place on 4 August 1911, it follows that Dawson's momentous visit was either on 10 May 1907 or 3 October 1904 (which would fit the vague *Geological Journal* version) or on 27 July 1899 (which would fit the far more explicitly stated date of the *Hastings Journal* and both Woodward's and Abbott's statements). Dawson had assumed his Stewardship by the previous year, but presided over the court for the first time in 1899. If the latter is the more likely date, as would appear from Abbott and Woodward's as well as Dawson's accounts, then the two labourers must have kept Dawson's request in mind, or been reminded of it, over a period of *some eight or nine years* before 'one of the men' (and the context makes clear, one of the *same* men) at last alighted on a piece of cranium. From his paper to the Geological Society, one would gather that the lapse of time was far less.

Writing in February 1913, Abbott tells the well known

'coconut' story—that the men found what they took to be a coconut and broke it up and threw the pieces on the gravel heap. In Abbott's version, Dawson himself 'after a great deal of searching recovered the first fragment'. In Woodward's *Earliest Englishman* version of the 'coconut' story, one of the men subsequently alighted on a piece of the skull they had partly shattered in ignorance of its true nature, and the piece was kept and handed to Dawson. The coconut story in this version is mentioned in the Dawson obituaries and is told by Miss Kenward, who believes that her father was apprised of the event by the workmen themselves. But Dawson writes nothing explicitly of a 'coconut'; he even seems to contradict it in his surmise 'that when the workmen first dug up the skull it was complete in most of its details, and that it was shattered and mixed with the gravel *before any part of it was noticed by them*'[4] (my italics). Thus the origin of the coconut story sinks into obscurity.

If Dawson has left the history of the early events quite uncertain, the disregard for dates is matched by the casualness of the actual excavation. We know that there was a succession of different things uncovered later in 1911, in 1912, and in 1913. Apart from eoliths, four pieces of cranium, a piece of jaw, five remnants of animal teeth, two flint tools (twelve objects in 1912), a canine tooth, four animal teeth, three flints (eight objects in 1913), in all about twenty important pieces of spoil, came to light in these two years of 'systematic' digging. But the excavators must have been anything but methodical, for there is an entire absence of exact measurements or of any plan or record of where the different finds were made in relation to one another. It is true that most of the objects were found in gravel previously disturbed by the labourers but several of the most critical items were stated expressly as found *in situ*. That a working plan to show the disposition of the different finds inside and outside the pit could have been made goes without saying, for such a plan was drawn long after-wards by Osborn[5] from information supplied by Woodward in 1920.

In one respect the work was done systematically at least in the 1913 season.

> All the gravel [writes Dawson] *in situ*, excavated within a radius of 5 yards of the spot where the mandible was found, was set apart and searched with special care. The gravel was then either washed with a sieve, or strewn on specially-prepared ground for the rain to wash it; after which the layer thus spread was mapped out in squares, and minutely examined section by section.

When Woodward states[6] that 'part of the lower jaw and the lower canine tooth were eventually found in the adjacent *undisturbed* [my italics] gravel', one should of course realize that the gravel had been removed, and 'in the spread' Father Teilhard found the canine.

Nothing is more crucial than the provenance of the small piece of occipital bone, found by Smith Woodward himself, in relation to that of the jaw. Woodward[7] informs us that the bit of occipital was found by him 'in another heap'—that is to say, from the context, another 'heap of soft material rejected by the workmen'; whereas of the mandible he writes, 'this Mr. Dawson dug out' while 'exploring some untouched remnants of the original gravel at the bottom of the pit'. The latter tallies with Dawson's own statement of 18 December 1912,[8] that the jaw came from 'undisturbed gravel'. He goes on to say that 'Dr. Woodward also dug up a small portion of the occipital bone

. . . from within a yard of the point where the jaw was discovered, and at precisely the same level', but he leaves unclear the critical question whether the occiput was in disturbed gravel or not. This can be judged from the paragraph as a whole:

> Apparently the whole or greater portion of the human skull had been shattered by the workmen, who had thrown away the pieces unnoticed. Of these we recovered, from the spoil-heaps, as many fragments as possible. In a somewhat deeper depression of the undisturbed gravel I found the right half of a human mandible. So far as I could judge, guiding myself by the position of a tree 3–4 yards away, the spot was identical with that upon which the men were at work when the first portion of the cranium was found several years ago. Dr. Woodward also dug up a small portion of the occipital bone of the skull from within a yard of the point where the jaw was discovered, and at precisely the same level.

Thus, though he speaks of broken fragments coming from disturbed gravel, we cannot be sure that he meant it to apply to the all-important occipital fragment.

Dawson provides another version in the *Hastings Naturalist* of 25 March 1913 which is quite unequivocal.

> For the most part [he writes] our work consisted in sifting the debris left by the former workmen, with occasional excavation into undisturbed patches of gravel which had been overlooked by them. In one such patch Dr. Woodward found a piece of the cranium bordered with a portion of the lambdoid suture within a yard of where the piece of jaw was found.

Woodward was fully aware of this paper by Dawson, as he had a signed (and dated) reprint in his collection.[9] His 1948 account thus flatly contradicts Dawson's. As he was the finder of the small piece and his statement even goes to weaken his own belief in the association of jaw and cranium, one must prefer his version. It is Dawson who is insisting on the *in situ* provenance of these so vital pieces of bone. But it is surely amazing that in the scientific paper to the Geological Society this utterly crucial matter is left so vague and equivocal.

More could be said of the rough-and-ready indications given to the locations of the various finds. Perhaps the photograph taken on 12 July 1913 of the Geologists' Association excursion to Piltdown (Pl. 8) may serve as a not unfair portrayal of the conduct of operations. For, swarming on the site on that sunny day, all round and right inside the pit, are to be seen the visitors. The significance of this poor archaeology lies in this: that Dawson was not an inexperienced excavator; that in the 1890s, as we know, he had undertaken what seems to have been an extensive dig in the Lavant Caves; and that in 1907 he had been in Mr. Ray's company when that archaeologist excavated, 'with photographs and measurements', two Neolithic skeletons on the Duke of Devonshire's estate.

Other points of interest emerge from a comparison of the several published papers. The flint tools found and reported in December 1912 (and figured in the April 1913 issue of the *Journal of the Geological Society*) number not more than *three*, but Dawson speaks of them in several places as if they had been found in considerable numbers. 'In the *majority*', he writes,[10] 'the work appears chiefly on one side, &c.' and he goes on to give the sketchiest descriptions of their workmanship. In describing the fragmentary mandible, Dawson[11] mentions that the broken-off condyle (the tell-tale articulation with the mandible) has 'rotted away', though in the scientific version nothing is said specifically on this point. But even a cursory examination of the part disposes of this suggestion, for there is very definitely a break and not an erosion. His inaccuracy is particularly evident when writing of the nasal bones which he himself had found in 1913. In the 'official' paper he himself mentions that under the nasal bones the turbinal was dug up in so bad a condition that it (the turbinal) fell apart. The nasal bones Woodward mentions as extremely well preserved (as they are). Dawson, in the second paper in the *Hastings Naturalist* (June 1915), now says: 'I was fortunate in finding the right and left nasal bones *in situ*, and though they crumbled to fragments, they have now been pieced together. They are short, &c. . . .' These last points are not trivial, for they are sufficient to make us wonder why Woodward, Teilhard, and Keith found Dawson meticulous, methodical, painstaking, and so on. In this second paper Dawson presents a description of the finding of the bone implement which one can only describe as misleading. From it the reader might well think that the implement was found in a mud-bank immediately under the *Eoanthropus* bed;[12] 'where it had been broken into many pieces' by the excavator. What Dawson does not say is that the implement was uncovered not in the gravel at all, but 'in dark vegetable soil beneath part of the hedge which Mr. Kenward had allowed us to remove',[13] and in that spot two large pieces, which fitted together, came to light. They inferred that it came from below the gravel because it agreed in appearance with some small fragments of bone found actually in place, and was smeared with clay of a type found at that level. But Dr. Oakley has given reasons for regarding the *in situ* material as spurious.

The problem of establishing what in fact Dawson brought to Woodward at the Museum in May 1912 has already been under consideration. We saw that some definite support for Woodward's list of the things, which included five cranial fragments, could be adduced. What of Dawson's version? Curiously enough, Dawson nowhere in the published accounts admits to the possession of more than *two* pieces of the cranium at a time vaguely specified as just before his visit to Woodward. In the first Geological Society paper he presents his history of only two fragments, one of 1908 (presumably), the other of autumn 1911, and then says merely, and by no means logically: 'As I had examined a cast of the Heidelberg jaw, it occurred to me that the proportions of this skull were similar to those of that specimen. I accordingly took it [*sic*] to Dr. A. Smith Woodward.' The Hastings paper has a more graphic description of that visit, but, as before we only hear of two fragments. In neither place does he admit to presenting Woodward with rather more spoil—nothing of the hippopotamus or stegodon teeth or the flints. But in two of his letters we possess substantial confirmation of Woodward's statements: on 28 March 1912 he acknowledged Woodward's 'determination of the tooth of hippopotamus' which he had sent on the 26th, and on 23 May he offered to bring 'the piece of skull and a few odds and ends found with it, or near it, in the gravel bed'. Publicly, Dawson nowhere admitted to bringing flints (to us suspect both in patina and workmanship) and hippo teeth (one definitely fraudulently stained). These tools and the tooth prove that when Dawson visited Woodward the forgery or hoax was already in the making. The hippo premolar, like the molar, has the artificially applied staining—chromium and iron. Its provenance, as we have seen, cannot be Piltdown. 'Stegodon', in Woodward's hands at the beginning, repeats the same tale. It is chromate-stained and its provenance on the basis of its radio-activity cannot be ascribed to any deposit in Europe. Whichever of the flint tools (apart from that found *in situ* later by Teilhard) was brought, it, too, would have been a falsification.

In fact, everything which Woodward saw in Dawson's hands that May (with the exception of the unimportant eoliths) was a falsification.

To clear up once and for all Woodward's position in the matter at the time, we need only emphasize that Dawson's letters of February 1912 (like his account in the two scientific reports) make it absolutely certain that Woodward knew nothing of the Piltdown finds before that time, despite the fact that he had been in touch quite regularly with Dawson during the previous three years or so. Before 1909, Dawson's contacts with him were only desultory; there is only one letter in 1903, a request for Woodward to look at some material discovered by Edgar Willet; then in 1907 a single friendly chatty letter regretting Woodward's failure to come to some party or gathering at Castle Lodge (perhaps a house-warming). In March 1908 (about the time of the first appearance of Piltdown man), the correspondence becomes regular and takes on a definite theme. It is initiated by Dawson with a recountal of his present interests—his explorations for the Heathfield gas, and his activities on the Hastings Museum Committee. He also makes mention of some reptilian bones that might possibly interest Woodward. In succeeding letters several novel ideas and diverse finds are mentioned and the offers of *Iguanodon* bones are made explicitly, and after these have been dispatched there are offers of the smaller things—the Wealden teeth, which to Woodward's, Dawson's and Teilhard's delight turn out to be of special value. The hard work of finding these Wealden mammals, the detailed examination of the finds, arrangements for publishing, the question of their ultimate disposal—such topics continue throughout 1910 and 1911. But never is there any mention of the treasure from Piltdown, known already to Kenward, Woodhead, Teilhard, Butterfield, and Lewis Abbott.

The Wealden teeth were described to the Geologists in London on 22 March 1911. Then follows a break in the correspondence. In February 1912 comes the casual announcement: 'I have a portion of a skull which will rival *Heidelbergensis* in solidity.' Again, in 1915, as we can judge from Dawson's letter of January and his postcard of July, the announcement of the new finds of site II came 'out of the blue' to Woodward. Woodward was not implicated in the business at all. There is not the slightest question of his integrity.

Here we may also dispose of the notion that Woodward ever denied free access to the Piltdown originals. Before the meeting of December 1912, it is true, he was secretive about them; but in the next few years Elliot Smith, F. G. Parsons, Keith, Pycraft, Le Gros Clark (then a medical student), Underwood, and later on Osborn, Weinert, Hrdlička, and Theodore McCown were all granted unrestricted view of the relics. But it is unfortunately and understandably true (as Mr. Hinton has pointed out) that in their day-to-day study of the fragments, the workers in the British Museum used Barlow's casts. For all their excellence, these do not show up the details on the teeth at all faithfully.

We can regret that the originals themselves were not perhaps more closely studied; we can deplore the omissions and inaccuracies in Woodward's recountal of events and the difficulties he has left in the way of our understanding certain crucial matters, such as the time and circumstances of his learning of the chromium-staining; we can be critical of Woodward over his omission to have the jaw analysed[14] as the cranium had been; all in all, Woodward in the matter of Piltdown may not have come up to the high standard of thoroughness for which he was renowned; yet, whether or not we allow him the extenuation that he was an immensely busy scientific worker with multitudinous responsibilities and that he nevertheless spent an inordinate proportion of his time on Piltdown and devoted much of his years in retirement to it, it remains quite certain that he had no hand in the great hoax.

Woodward himself had only the smallest qualms, as far as we judge, about *Eoanthropus dawsoni* till the end of his days. He recognized some difficulties in the canine. He was disappointed rather than worried over the exact location of the second site, and certainly attributed Dawson's vagueness to his last long illness.[15] He was quite furious when Hrdlička voiced a doubt on the existence of the second site and sent the then Keeper of Geology, Dr. Lang, 'an important postcard from the late Charles Dawson, 30 July 1915, in which he announces his discovery of the isolated lower molar tooth of *Eoanthropus, with the remains of the second skull* in the Piltdown gravel . . . there may be doubters for whom Dawson's own record is needed'. Manifestly, Woodward himself was no doubter in view of the statement italicized (my italics). On a visit to Piltdown with him, in his retirement, Professor van Straelen of Brussels[16] remembers Woodward's 'expressing cautious opinions' and not 'feeling comfortable' about the jaw and tooth. Professor Swinnerton knew Woodward very well personally and he gained the impression that Woodward himself 'had streaks of doubt about the jaw'. But clearly such doubts would refer to its association with the skull and not to its authenticity

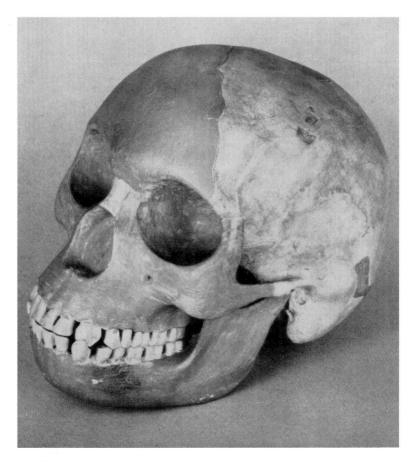

1. **The Dawn Man of Sussex**—*Eoanthropus dawsoni*
    (A composite reconstruction of brain-case and jaw)

2. The gravel pit at Barkham Manor, Piltdown, Fletching, Sussex. The lowest layer is the Tunbridge Wells sands and above it are the gravel deposits in which the specimens were found.

3. **Personalities concerned with the Piltdown discovery.**
Back row: Mr F. O. Barlow, Prof. G. Elliot Smith, Mr C. Dawson, and Dr Arthur Smith Woodward.
Front row: Dr A. S. Underwood, Prof. Arthur Keith, Mr W. P. Pycraft, and Sir Ray Lankester.
(*From the Portrait painted by John Cooke, R.A., in 1915*)

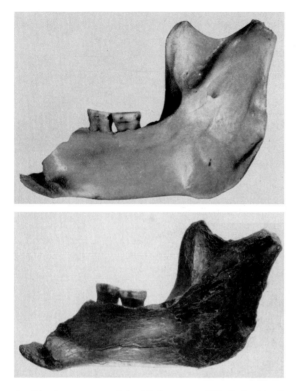

4. The inner aspect of the Piltdown mandible (below) is shown for comparison with a mandible from a female orang-utan (above); the upper specimen has been broken and the teeth abraded to simulate the Piltdown specimen. (*Prepared by L. E. Parsons*)

5. The cast of the Piltdown canine (left) is compared with a canine from a rather more mature orang-utan; the latter tooth has been stained and abraded to simulate the Piltdown specimen.

6. Details of the worked
end of the Piltdown 'bone
implement' (below) for
comparison with a fossil
bone from Swanscombe
(above), whittled with a
steel knife by Dr K. Oakley
to show that the Piltdown
fossil bone could only have
been worked in this way in
modern times.

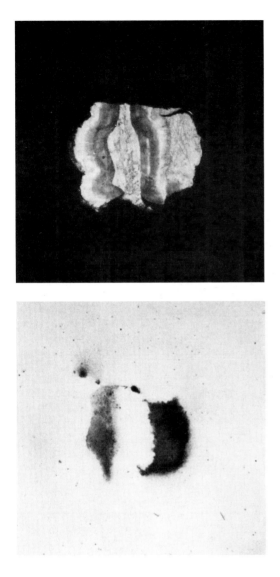

7. 'Stegodon' from Piltdown
   Above is a section of a fragment of one of the Piltdown elephant teeth;
   below is the auto-radiograph obtained by contact with a very sensitive film.
   It reveals the extraordinarily high radio-activity of the dentine and cementum
   layers—an intensity not present in any of the other Piltdown specimens, or
   in any comparable specimens from Britain or Western Europe.

8. The excursion of the Geologists' Association to Piltdown on 12 July 1913.

9. **The Piltdown flints (palaeoliths).**
The third specimen from the left is E.606, the staining on which contains chromium. The fourth specimen with inscriptions came from the collection of the late Mr Harry Morris of Lewes.

as a fossil. In his talks to Mr. Edmunds in 1926 and Dr. Leakey in his last years his confidence seemed entirely unruffled.

---

1. *The Earliest Englishman*, 1948, p. 10.

2. Dawson sent Woodward a notice of the Court Baron for 1911 saying (25 May 1915) that 'it was one of those four yearly meetings which led years ago to the gravel bed at Piltdown'.

3. I am indebted to Sir Percy Maryon-Wilson, Lord of the Manor of Barkham, for permission to inspect these records and to Mr. Steer, F.S.A., County Archivist at Lewes, for help in tracing the information.

4. Dawson, 1913, op. cit., p. 77.

5. Op. cit., p. 56.

6. 1915, op. cit., p. 9.

7. 1948, op. cit., p. 10.

8. Dawson and Woodward, 1913, op. cit., p. 121.

9. Now in the Woodward Collection of Zoological Books and Reprints in the Library of University College, London.

10. Dawson and Woodward, 1913, p. 122.

11. Dawson, 1913, p. 80.

12. Dawson, 1915, p. 184.

13. Woodward, 1948, p. 12.

14. Or at least mentioning that it was unfortunately too difficult with the small quantity of material that would have been available.

15. Personal communication from Lady Smith Woodward.

16. Letter to Lady Smith Woodward, Nov. 1953.

11

Entanglement

⌑ ⌑ ⌑

Remove you skull from out the scattered heaps:
Is that a Temple where a God may dwell?
Why ev'n the worm at last disdains her shattered
cell!

Look on its broken arch, its ruined wall,
Its chambers desolate, and portals foul:
Yes, this was once Ambition's airy hall . . .
BYRON: *Childe Harold's Pilgrimage.*

To Professor Teilhard de Chardin the idea that either Dawson or Woodward was in any way wittingly implicated in this business is completely unthinkable. He holds both of these men in the greatest respect, and indeed is inclined to doubt whether a real hoax occurred at all. It seems to him not impossible that the pit at Barkham Manor was used as a rubbish dump where, over a course of years, all sorts of objects including bones from some discarded collection could have been deposited. The heavily iron-bearing water of the gravel would soon stain the bones dark brown (for Professor Teilhard says that fresh bone left in the water of the Weald does stain easily). To be sure, the queer accumulation must have come from some collector's hoard! But this suggestion of a rubbish dump leaves, of course, too much unexplained and far too many coincidences. It would be an amazing accident that would bring together an unusual cranium, the jaw, the remarkable canine, a bone implement of unique character, a number of flints of spurious workmanship (one of which is stained with chromate), and bones partly changed to gypsum and radio-active fossil teeth of a sort never found in England! That all this curious medley, this 'accidental' assemblage, should be uncovered in a particular sequence—as Sir Arthur Keith said to us, 'as if to confute me personally!'—is straining our acceptance of coincidence too much.

Miss Kenward, who lived at Barkham Manor for many years, is positive that the pit was not a general rubbish dump and that the gravel was being dug from an unbroken surface. Lady Smith Woodward, too, rules out the suggestion. There is nothing to commend this rubbish dump theory, for we have seen that hardly anything of the whole collection of material can with certainty be said to have come from gravel originally, although this does not rule out the possibility that at least the cranium, even stained as it is, may not have been genuinely found in the pit and that after treatment it was redeposited in the gravel pit. This we shall have to consider in due course.

We have learnt that in treating of the cardinal events at Barkham Manor, in dealing with the dates of the first discovery of the gravel bed, and of the first cranial piece, the provenance and particulars of the objects brought to Woodward, the crucial question of which finds were actually *in situ*, and the exact location of the bone implement, Dawson's own statements fall far short of establishing beyond question the validity of most of the basic documentary evidence.

The credibility of the early finds was in fact called into question publicly. Mr. A. W. Oke (an amateur member of the Geologists' Association, the Sussex Archaeological, the Hastings Naturalist, and other societies) wrote an extremely hostile letter to a Brighton paper in 1926 in this vein. As Woodward had a cutting of this letter in his possession, he was well aware that Dawson had his detractors, and as a member, and in the last years President, of the Sussex Archaeological Society, Woodward was in touch with members of its Committee, who, we know, were sceptical of the discovery from the start. But Woodward, like Keith, had complete faith in his colleague. As Keith wrote:[1] 'He and I never differed as to the genuineness and importance of the discovery made at Piltdown, and we had both the same love and respect for Charles Dawson, the lawyer antiquarian. . . .' His confidence in Dawson never wavered and he paid his tribute to his colleague in the memorial stone at Barkham Manor.

Clearly, for Woodward there was nothing untoward about the second discovery at Sheffield Park in 1915. He and Dawson had explored the neighbourhood as a possible site in the spring or autumn of 1914, while Teilhard had been conducted there by Dawson in August 1913. The inadequacy of the information about the exact site was in Woodward's view attributable to the unexpected course of Dawson's illness, which began in the autumn of 1915. Likewise, Mr. E. V. Clark, who saw him on Lewes Station quite often, always hale and hearty, remembers how insidious was Dawson's last illness. It is not surprising, therefore, that Woodward did not trouble the sick man for the information. He did not visit him during that time (says Lady Smith Woodward), but he sent him a message in June 1916 to cheer him with the news of another season's digging. By the time Woodward tried to get the particulars he wanted, Dawson was too ill to supply the information. Mrs. Dawson told a friend what a tragedy it was that her husband had become too ill to tell her clearly about a skull to which he kept alluding in those last days.

The second Piltdown discovery of which we know so little is a repetition of that at Barkham Manor as a tale of fabricated rarities, of insecure provenance, found in ill-defined circumstances.

The Piltdown II fragments are fraudulent or deceptive in no less than four different ways. The cranial bones do not belong to the tooth; the frontal and occipital fragments have been stained artificially with both iron and chromium; the fragments were associated in an undisclosed way with a Red Crag rhinoceros tooth whose presence in that situation is quite impossible to accept; finally, the two fragments do not belong to the same skull—in fact, the frontal piece must come from Piltdown I. All these things were delivered by Dawson to Woodward, thus duplicating the discoveries of 1912—a man-ape combination with a Red Crag tooth to establish the near-Pliocene date. It also duplicates the staining techniques. To effect all this, the perpetrator retained for some years a piece of the original skull and had available pieces of other human skulls as well. He used exactly the same technique as before when Piltdown I was 'salted'.

'I believe we are in luck again', wrote Dawson on 20 January 1915 to Woodward. 'I have got a fragment of the left [sic] side of a frontal bone with portion of the orbit and root of nose.' And in July 1915: 'I have got a new molar (*Eoanthropus*) with the new series.' Woodward gives early in 1915 as the date of the finds of the three pieces of the second Piltdown material and adds: 'shortly afterwards, in the same gravel, a friend met with the piece of rhinoceros molar'.

It thus appears from the letters quoted and the scientific reports that the four objects from site II were reported to Woodward separately—the piece of frontal in January 1915, the molar in July, but when the occipital fragment and the rhinoceros tooth were announced we do not know, nor do we have the name of the 'friend'.

It is surprising that Woodward failed to obtain fuller details before Dawson's illness began in October, for Dawson was certainly in touch with him till then; on 20 January he had written to say he would bring the new-found frontal bone soon, and he was in London to give his 'anti-eolith' paper to the Royal Anthropological Institute on 26 February. They may have met at the sittings for the Royal Academy portrait. Dawson's professional duties prevented him, however, from attending the Academy Private View on 30 April, nor was he able to go to a demonstration by Keith at the Royal College of Surgeons in July.[2]

On the face of it, these finds at site II are decidedly odd. That they came off an unspecified ploughed field, 'when in the course of farming the stones had been raked off the ground and brought together in heaps', as Woodward wrote, is very strange when we consider the small size of the fragments, especially the molar, and the fact that they were presented to Woodward as found at different times.

Did Dawson withhold the details of these finds because at the time they came into his possession, or even long before, he had discovered that he had been fooled? Was he in 1915 painfully aware for certain that some of the things he had originally taken to Woodward were forged, as was much else in the succeeding years?

This is what many people, including Mr. Francis Vere, who broadcast in defence of Dawson, think happened to Dawson, and why at the end he never told Woodward the exact site. And it would perhaps be possible to surmise that suspicion aroused as far back as the first public announcement would account for the unsatisfactory way in which Dawson reported his discoveries when he finally came to describe them for the Geological Society paper of March 1913. Alternatively, if we are to dismiss this sinister interpretation, we are bound to regard Dawson as observer and recorder of almost culpable inadequacy.

As we ponder the curiosities and inconsistencies of the recorded and accessible history of Dawson and his discoveries there emerges inexorably the question: 'Was he guiltless and oblivious of it all, enthusiastic yet grossly careless and unsystematic, or was he indeed knowingly involved in the great deception in some way?'

Apart from the unsatisfactory way in which Dawson reported the finds at Piltdown II, was he aware of the artificiality of the assemblage there? In his letter of January 1915 he gives a perfectly adequate description of the new frontal:

> I have got a fragment of the left side of a frontal bone with portion of the orbit and root of nose. Its outline is nearly the same as your original restoration and being another individual the difference is very slight.
>
> There is no supra orbital foramen and hardly any superciliary ridge.
>
> The orbital border ends abruptly in the centre with a sort of tubercle and between it and the face is a groove or depression ¾ inch in

length. The section is just like Pycraft's model section and there are indications of a parietal suture. . . . The tables are thin—diploe very thick. The general thickness seems to me to correspond to the right parietal of *Eoanthropus*.

He is quite positive this is another individual because, like that of Piltdown I, he decided that it was a *left* frontal, despite his close examination. It is in fact from the right side. Woodward would no doubt have corrected him on this when he brought the piece along, as he said he would, in January. Yet there was something like prescience in his mistaken diagnosis, for the occipital which he found at some other unrecorded time certainly is of a second individual. As we know, it cannot belong anatomically or chemically to Piltdown I, but neither can it belong to Piltdown II! The hoaxer was determined to have his second Piltdown man—the complete answer to silence the sceptics. The indubitable second occipital made it impossible to believe that the frontal or the molar could, by mistake or by removal of gravel, have come from the first site, two miles away. This episode, which established the second man, is one of a number of striking examples where Dawson's successive actions seem uncannily to mirror the developing plan of the perpetrator. The chromium-staining is another of such instances.

This question of the significance of the chromate-staining present in both places posed by the Piltdown II remains has been raised acutely already by Piltdown I. This is undoubtedly a most serious aspect of the affair. For there can be little doubt that chromate represents a part of the deliberate staining which has gone into the deception. The evidence has been presented in some detail. Briefly, the hippo and 'Stegodon' teeth, known to be importations (on other grounds), were stained with iron and chromium, as was the jaw dug up in Woodward's presence, as were the first cranial pieces.

What has Dawson's reported use of bichromate on the first pieces of cranium to do with the illegitimate chromium-staining? Would Dawson have made the admission had he known the jaw had been stained, for the staining on the jaw could only mean to him what it means to us? And knowing so much he would have already been in the secret of the hoax. If he did know something more of the staining, why tell Woodward at all? It seems a great risk for him to take, and there seems no obvious reason in the coloration itself to force Woodward's curiosity. It is true that the cranial bones vary somewhat in their brownish colour and the temporal bone is conspicuously a much more yellowish colour than the rest. But iron-staining could be expected to vary quite naturally in a series of fragments from such iron-bearing gravel beds.

If we accept that Dawson personally told Woodward of this staining, there seem three possible explanations. One possibility is that the chromium was used innocently by Dawson on many more of the 'finds', but that having been reproved by Woodward, who said it was based on a mistaken notion, he felt chary of admitting to the pointless preservation of such solid fossils. This raises some serious difficulties. It implies that some of the objects, like the jaw, found in Woodward's presence, were entrusted to Dawson for a period, and that he 'hardened' them in the bichromate solution, whereas he did not do this for the piece of occiput, for example, dug up by Woodward. Lady Smith Woodward believes, however, that her husband took the finds including the jaw away at once, as he did the canine, and of course one can see no reason why he should not do so. Another difficulty is the bichromate-staining of the flint tool (No. 606). Surely there can be no question of 'hardening' this or any one of these three of four flints. It is altogether difficult to believe that Dawson could have carried out the treatment in so uneven a way for the innocent or mistaken purposes of preservation; and if he 'owned up' to one lot of staining, why did he not to all, as Woodward was obviously unperturbed. This explanation leads straight to attributing to Dawson the staining of the 'importations' and fabrications.

The second possibility is that when Woodward in some way came to inquire about the colour, Dawson had by then 'tumbled' to the whole deception and, knowing of the chromium, admitted to the stained condition only of the first cranial fragments. This explanation, too, means that Dawson 'was in the know'.

Lastly, we can regard his use of the chemical as wholly innocent, though pointless ('mistaken', as Woodward said), and restricted to the first cranial pieces. It thus therefore becomes a remarkable coincidence that the forger should have had occasion to use the same salt extensively.

It is unfortunate that we cannot be quite sure that Dawson himself told Woodward about the bichromate. As we know, Woodward mentioned the matter for the first time twenty years after Dawson's death. In that time he might have had the information directly or indirectly from other sources. Mr. Edmunds was told about the use of the bichromate in 1926 by Lewis Abbott, but does not recall passing this information on when he reported on the geological results of his survey to Woodward (who incidentally was not dissuaded by Mr. Edmunds' findings from his belief in the Early Pleistocene Man). Abbott's mention of the 'first' pieces suggests that he may, directly or indirectly, have been the informant.

Woodward's own words are not unequivocal as to whether he had the information personally from his fellow excavator: in 1935 and in 1948 he wrote: 'the colour of the pieces which were first discovered was altered a little by Mr. Dawson when he dipped them into a solution of bichromate of potash in the mistaken idea that this would harden them'. Professor Teilhard cannot remember anything being said of the bichromate or of its use at the time of the excavation, nor can Lady Smith Woodward or Dr. E. I. White, Smith Woodward's colleague. If Woodward's information came only after Dawson's death, it means that the extent of the latter's use of bichromate treatment becomes uncertain and was not necessarily confined to the first pieces. It becomes the more serious in its implications, as we have only Abbott's reported word that the first pieces only were stained, and the purpose of the staining would also be conjectural. Despite these uncertainties, we can, however, be quite sure that Dawson did possess this chemical, for he certainly used it in connection with his 'eolith' study.

For his paper to the Royal Anthropological Institute in February 1915, on the human or natural workmanship of eoliths, Dawson propounded a theory that the breakages of the flint were entirely analogous to the natural fractures which could be induced in a prismatic material, such as starch. He accordingly illustrated his lecture with such starch 'flints', and these he coloured to lend verisimilitude to the demonstration, as, of course, he made plain to the audience. The starch 'flints' were given to the British Museum after Dawson's death, and the stain used on them contains chromium. But there is much more to be said of the staining activities of Charles Dawson.

In 1917, Mrs. Dawson, after her husband's death, gave to the Museum the skull fragments known as the Barcombe Mills skull. Before the public auction at Castle Lodge she arranged for her husband's collection of skeletal material to be looked over by Smith Woodward. On 7 January she wrote to the Keeper of Geology: 'I have not yet come across pieces of skull answering to your description, but as I am putting everything of that nature into a cupboard you will have a wide assortment from which to choose.'

Of the exact provenance of these Barcombe Mills skull fragments we know nothing except that they were obtained from the river gravel of the Ouse at the low level at Barcombe Mills. The fragments were not described until a few years ago by Professor Ashley Montagu[3] of Philadelphia. There is no record that Dawson or Woodward ever claimed anything of special interest for these remains, but the late Dr. Broom[4] declared them to constitute yet a third Piltdown man. Professor Montagu's examination reveals the skull pieces as consisting of the remains of at least two individuals; there is no lower jaw and there is nothing exceptionable about the material morphologically. The fragments are indistinguishable from modern bones and the cranial pieces are not as thick as those of the original Piltdown skull. Their negligibly low fluorine content and their high nitrogen content show them to be almost certainly modern. What they have in common with Piltdown is a somewhat similar brown colour.

These fragments from Barcombe Mills have been subjected to the very remarkable chemical treatment, so remarkable that it was the staining which they underwent which threw so much light on the iron-staining used so extensively by the forger. These Barcombe Mills pieces have adhering to them some gravelly matrix mixed with a little soil, and this material is also of a dark brown colour. Like that of the bones, the colour is attributable to artificial iron-staining, and in both cases the iron salt used was an acid sulphate, which formed gypsum as a by-product. The matrix provides undeniable evidence that this iron-sulphate-staining was deliberate, for the matrix contains, in addition to iron oxide, traces of ammonium sulphate, never known to occur in nature and indicating without doubt that the staining was produced by the use of iron alum, which is ferric ammonium sulphate, and which, like the combination of chromate and ferrous salts, is an efficient and recognized method of depositing an iron salt. The artificiality of the gypsum in the bone is confirmed completely by the virtual lack of sulphate in the gravels and loam in the river terraces at Barcombe Mills. The higher terrace is an extension of that at Barkham Manor and there, too, the absence of sulphate contrasts with the sulphate in the cranial fragments (Piltdown I).

The gypsum in the Barcombe Mills and Piltdown material was revealed by the X-ray crystallographic analysis. By boiling bone with ferric ammonium sulphate, Drs. Claringbull and Hey at the British Museum (Natural History) obtained complete proof of its use in all these sulphate-bearing fragments, for they found that gypsum replacement in the bone apatite could be easily reproduced by this means.

One treatment involved in the staining, the bichromate, was certainly known to Dawson. His possession of the Barcombe Mills 'skull' invokes the *a priori* possibility of his knowledge also of the iron sulphate technique.

Why should Dawson retain those Barcombe Mills fragments? To judge by the chemical analyses, they represent three different individuals, and it is difficult to believe that they could have been obtained from or represented as a single burial. Are we to suppose that he had fallen into the forger's trap yet a third time? Or we can, of course, in accordance with our alternative theory, presume that in staining them himself this might merely have constituted another episode in the process of discovering the way by which he and Woodward had been duped. We know in fact that he did carry out extensive experimentation in the staining of bones after the first pieces of the cranium were reposing in Woodward's care.

We have reason to believe that some of Dawson's experiments into staining methods, on both bones and flints, took place about May 1913. Once again his activities appear to coincide with that of the perpetrator, but such staining does not necessarily prove that Dawson was primarily responsible for the treated Piltdown fragments. Curiously enough, those three men who in 1913 became aware of the artificially stained condition of some of the 'finds' or who knew of Dawson's interest in these staining techniques did draw the worst conclusions.

---

1. *Autobiography*, p. 654.
2. This may have been an exhibit of Mr. Harry Morris's collection.
3. Montagu, M. F. A., 1951, 'The Barcombe Mills Cranial Remains', *Amer. J. Phys. Anthrop*, **9**, pp. 417–26.
4. Broom, R., 1950, Summary of a Note on the Piltdown Skulls', *Adv. Sci.*, **24**, p. 344.

12

'The Eye Wink'

¤ ¤ ¤

In 1941 Mr. F. W. Thomas, on the staff of the *News Chronicle* and the *Star*, was advised to evacuate from Seaford and went to live in Lewes where he and his wife stayed with Mr. A. P. Pollard, Assistant Surveyor of the Sussex County Council. One day as they were touring round Chailey they found themselves near the famous Piltdown site. They had some discussion of the gravels and the circumstances of the finds. When his visitor remarked on the great evolutionary importance of the Piltdown man he was extremely surprised at his guide's reply, which was that there was really nothing in the great discovery, and that he was entirely sceptical of it all. Mr. Pollard did not add anything more at that time.

From Mr. Salzman (now President of the Sussex Archaeological Society) I learnt in August 1953 that Mr. Pollard was well acquainted with the gravels and gravel workings in the Lewes region, and that he might be able to help me with my inquiries on the history of Piltdown.

When I explained I was interested in the discovery, Mr. Pollard immediately asked me whether I had any reason to distrust the discovery, and on my admitting as much, he said, 'I am not surprised. I believe it to be a fraud. At least, that is what my old friend Harry Morris used to say.' What Mr. Pollard had to tell me he had learnt from Harry Morris, a bank clerk and keen amateur archaeologist, whose acquaintance he had first made on taking up his post at Lewes in 1928. Morris and he became close friends, and he was Morris's executor and saw to the donations of the latter's collection of eoliths and other flints to the two Lewes museums.

Morris in 1912 or 1913, right at the beginning, had come to the conclusion that the flints at Piltdown were not genuine. When he first saw the flints, he at once rejected them because 'Harry Morris knew every flint bed and gravel bed in the district'. Morris was convinced that the Piltdown implements could not have come from the Barkham Manor pit and that they were by no means as early as the Lower Pleistocene as Dawson gave out. He always insisted that if they were genuine artefacts, they were Neolithic and no earlier.

'What proof had Morris for his accusations?'

Mr. Pollard explained that Morris had somehow managed to obtain from Dawson one of the flints which were supposed to have come from Piltdown, and from an examination of this flint he had become utterly convinced that some deception had been carried out. As far as Mr. Pollard could say, Morris had never conveyed his suspicions to anyone else, except perhaps to his close friend, Major R. A. Marriott, D.S.O., another enthusiast for eoliths. But Morris had gone so far as to write down what he suspected. This flint of Morris's and the written statements about it had come into Pollard's possession, for Morris had left him a large cabinet of flints. They were of little interest to Pollard, and in 1948 he exchanged this cabinet with Frederick Wood of Ditchling for a collection of birds' eggs. Fred Wood had been dead some years and Pollard could not say what had become of the cabinet. Mrs. Frederick Wood still lived at Ditchling.

Some weeks later, with my colleague, Mr. Geoffrey Harrison, I was able to go down to Ditchling and found the cabinet was still there. In the years since her husband's death the specimens had not been of interest to anyone, and we feared that the evidence would have gone. There were twelve drawers, nearly all filled with flints of various kinds, though mostly of Harry Morris's easily recognizable eoliths, everything neatly labelled. We started with the top drawer and worked our way down, our anxiety increasing as we proceeded. In the twelfth and last drawer of all was the 'Piltdown' flint, and with it two documents.

This flint (Pl. 9) is now preserved in the British Museum (Natural History). It is somewhat quadrangular in outline, with a flat base; one area has been heavily battered before the core was flaked. The flint is basically grey with inclusions, identical in lithology with the three 'pre-Chellean' Piltdown flints. It also has a similar patina and staining. On the flint, in Morris's writing, appear these words:

> *Stained by C. Dawson with intent to defraud (all).—H.M.*

In the accompanying note Morris has written:

> *Stained with permanganate of potash and exchanged by D. for my most valued specimen!—H.M.*

On the second card, which is a piece of cardboard photographic backing, we read:

> I challenge the S[outh] K[ensington] Musuem authorities to test the implements of the same patina as this stone which the impostor

Dawson says were 'excavated from the Pit~!' They will be found [to] be *white* if hydrochlorate [*sic*] acid be applied.

H. M. Truth will out.

As Morris claimed, the application of dilute hydrochloric acid to the surfaces of the three Piltdown 'palaeoliths' as well as to the 'Morris' flint does dissolve off the orange or yellowish brown stain, leaving a pale yellowish or greyish white surface. In marked contrast are the 'eoliths' and other flints collected from the Piltdown gravel. Their brown patina is unaffected by the acid. The staining of the palaeoliths is surprisingly superficial; below its surface the cortex proved to be pure white. Yet in the brown flints normally found in the Piltdown gravel the cortex is iron-stained throughout its thickness. Flint nodules with white cortex do not occur naturally in the Piltdown district.

Harry Morris was correct in alleging that the staining of the flints was deliberate. The iron-staining is in all likelihood artificial. But Morris was quite wrong in suggesting that permanganate had been used, for no manganese can be detected on any of the flints. But the reader will recollect that one of the flints carries its own special evidence of fraud. This is the implement recorded as found by Teilhard de Chardin *in situ*; it has the same thin removable iron, but in addition it contains appreciable traces of chromium. No chromium can be detected in the gravels of Barkham Manor.

Of Mr. Pollard's intimation to me that Morris considered the flints of Neolithic date, Dr. Oakley was unaware when he judged independently that they could best be matched (apart from their colour) 'in the flint debitage found at flint-mining or chipping sites of Neolithic or later age on the Chalk Downs of Sussex'. Dawson was indeed not far from the truth when he wrote of the Piltdown palaeoliths that 'they resemble certain rude implements occasionally found on the Chalk Downs near Lewes, which are not iron-stained'.

Morris did not confine his portentous accusation to the deception of the flint. On the second note he records this: 'Judging from an overheard conversation, there is every reason to suppose that the "canine tooth" *found at P. Down was imported from France.*' (Morris's italics.) That is all, except for the cryptic and melodramatic words in pencil across the ink-written accusation: '*Watch C. Dawson. Kind regards.*' Had Morris intended to send this curious document and changed his mind? How he came to write down these accusations we do not know. But that he had conceived an intense antipathy towards Dawson amounting to hatred is abundantly certain.

A strong element of resentment and perhaps jealousy entered into Morris's attitude. In the days of Dawson's triumph, Morris was meeting with much opposition to his claims for 'eoliths'. He is well remembered as a fanatical believer in the human workmanship of his own brand of 'eolith', but he got little support even from the doyen, old Benjamin Harrison, or from Reid Moir, though Keith encouraged him. His claims were examined in 1913 by a committee of geologists, who reported very unfavourably. Though I have not traced the records of this committee, Morris replied to the many criticisms with all the arguments and irony he could command; the newspaper report of this has been preserved among Morris's press cuttings, along with various uncomplimentary jottings on 'armchair critics', & c.

Morris became a man with a grievance. Arthur Keith listened to him sympathetically, and retains a memory of the bachelor bank clerk of Lewes. Keith noted in his journal on the occasion of his visit in July 1913 as one of the party of about 100 members of the Geologists' Association invited by Dawson to visit the Sussex sites:

> In Lewes I was besieged by a Captain Marriott: I had been in correspondence with him before and he urged upon me the merits of his friend, Morris. We went to Morris's small lodging—two rooms—crammed to overflowing with eoliths and some palaeoliths. Morris is turning sour because of scepticism.

And Sir Arthur Keith adds:[1] 'I don't remember any mention of Dawson by him or Marriott, but I remember very well my own thoughts about Dawson's acclamation and poor Morris's neglect.' Keith did more at the time, for Morris was able to demonstrate his collections at the Royal College of Surgeons in 1915, at about the time when Dawson made his sharp attack on 'eoliths' at the Royal Anthropological Institute. Dawson's and Morris's opposing views were the subject of an annotation in the *Lancet*. Dawson's paper, we know, was never published, but some notes of Morris's indicate that he must have attended the meeting, for his mortification and resentment at the criticisms are plain. To the end of his days he stuck grimly to his 'eoliths', though it was a losing battle against archaeological opinion. He came to Oxford to give a paper to the University Archaeological Society in 1920, and there had a distinguished but unsympathetic audience. Sir Arthur Evans, Mr. Henry Balfour, and others voiced adverse comments, and Professor Sollas was outspokenly critical. Dr. S. Spokes, who argued in favour of Morris's contentions, had his home in Lewes, and he knew Morris as one of the local archaeologists; the Piltdown casts had been given by him to the Barbican Museum; he had known Dawson and Woodward and, as a dentist, had taken a keen interest in it all. At Oxford, Morris voiced to his son Mr. Percy Spokes (who had arranged the meeting) a scepticism of Dawson's discoveries, but said nothing definite.

Morris never sought to discredit Piltdown Man or Dawson in public. He was palpably caught in a dilemma. He nourished (in private) these serious accusations of fraudulent dealing, and we may be sure he would ardently wish to see the faked palaeoliths swept away in favour of his eoliths—the eoliths which Dawson had declared to be of no account. But to denounce the flint implements must bring the whole of the Piltdown remains into disrepute. And the continued existence of Piltdown man was vital to Morris. For no better evidence of the human workmanship and the genuine antiquity of the 'eolith' could be imagined than Piltdown man himself—the maker of eoliths, if only the palaeoliths could be discounted (as they deserved to be!). And this is the line Morris tried to take. In his Oxford talk and later in a printed note to an exhibit of his collections in 1929[2] he simply ignored the palaeoliths. But recognition never came as it had in some measure come to those other champions of Pliocene tool-markers, Benjamin Harrison and Reid Moir. As the years went on, the deeply disappointed Morris appeared, except to a few friends like Pollard and Marriott, a cranky and heterodox protagonist of a half-forgotten theory. Of Dawson's 'base ambition' he spoke often to Pollard. His image of the man he apostrophized in these lines from *Macbeth*, written on one of the accusing notes, under the title of 'Dawson's Farce':

> Let not light see my black and deep desires.
> The eye wink at the hand; yet let that be
> Which the Eye fears when it is done—to see!

Whether Morris ever himself attempted to convey his allegations to Smith Woodward we cannot say, and it seems unlikely. Morris's views would probably have been discounted on personal grounds. But a suspicion of fraud did come to be entertained at South Kensington by at least one of the palaeontologists there. The source of this would in all probability have been Major Marriott, the friend of Morris, and a very remarkable man. In the Piltdown days he was Governor of Lewes Jail.

Major Marriott had retired from the Royal Marine Artillery after a career which for its feats of daring brings to life the Victorian sagas of Henty. He served in H.M.S. *Monarch* at Tunis as Secretary to the Sfax Commission. He was at the Consulate at Alexandria in 1882 at the time of Arabi Pasha's rebellion and the massacres of Europeans. Before the Bombardment of Alexandria he went out in stoker's clothes, posing as an Italian seaman, and inspected the Egyptian guns in the Mex forts. For his part with the Royal Marine Battalion at Kassassine and Tel-el-Kebir he was mentioned in dispatches. Marriott organized and commanded the famous Egyptian Camel Corps in the Nile expedition of 1885–6. He was awarded the D.S.O. by Queen Victoria on a first occasion of the investiture. He entered the Navel Intelligence Department when it was first formed in 1887. A man of strong character, his independence of mind encompassed many unorthodox views. He had his own theory of the formation of the Weald, collected flints, and was a staunch champion of Morris's eoliths. In 1913 he joined the Geologists' Association, of which A. S. Kennard was a prominent member. On the occasions of his visits to London he would call on Martin Hinton, a friend and co-worker of Kennard, at the Natural History Museum. Major Marriott died in 1930.

Kennard, one of the 'Ightham' palaeontologists,[3] let it be known on several occasions (in the 1940s) that he believed Piltdown man to be a hoax. These utterances were made in a mock-serious manner characteristic of Kennard; they were not supported by any evidence, and they were never taken as other than fanciful ideas. An obituarist[4] said of him: 'Living to a ripe age, he became a valuable link with the geological workers of past generations and, as a humorist, he was not above some harmless gossip about those personal peculiarities which make people real.' If Kennard's information came from Marriott (or a closely allied source), as in all likelihood it did, and startling as it certainly must have appeared, it might well have left him not entirely convinced. At least he obviously did not feel so strongly as to pursue the matter seriously.

The information could have come to Kennard as far back as 1915 or even a year before. One wonders whether he and Reginald Smith were not indulging in some thinly-veiled irony when they offered their opinions (already referred to) on the newly reported bone implement at the meeting of December 1914. Kennard remarked that he wished to congratulate the authors [i.e. Dawson and Woodward] on the discovery of a new problem from Piltdown. From the differences between the cut portion of the bone and the natural surface, he considered it possible that the bone was not in a fresh state when cut. . . .

Reginald Smith, I am assured by his friend, Major Wade, was at that time in a sceptical frame of mind over the geology of the Piltdown gravel. Of the 'club'-like implement of *Eoanthropus* he remarked: 'The possibility of the bone having been found and whittled in recent times must be considered. . . .

Experiment might prove whether a similar surface could now be produced by cutting, as opposed to fracture. . . . The discoverers were to be congratulated on providing a new and interesting problem, such as would eventually provoke an ingenious solution.'

Of this discussion, Mr. Alfred Oke wrote in the hostile letter (to the Brighton paper) already referred to:

> I was present at the meeting of the Geological Society when Mr. Dawson produced a bone implement from Piltdown, which, he said, was found in the soil from the pits, but he had to leave to catch his train before he could be cross-examined.

The tenor of the letter can be gauged from another assertion referring to Piltdown man:

> The fragments of bone are only held together by the story of the workman bringing to the late Mr. C. Dawson only a fragment of what he and his mates thought to be a coconut. Mr. Dawson was a Coroner, and, therefore, understood the laws of evidence, but no Sussex jury would have been satisfied that the cleverly reconstructed skull consisted of bones belonging to the same being.

Piltdown man, for Major Marriott, was a fraud. It was well known in his family that he held this belief, though he spoke of it very seldom, as his daughter (Mrs. Olivia Lake) informed us after the disclosure of November 1953. Mrs. Lake recalls particularly that a picture in a newspaper of the Piltdown man was the occasion of her father's telling her that the jaw was faked, and so was the canine tooth. This is reminiscent of Harry Morris's accusations, and Marriott, we know, was in Morris's confidence. But it seems altogether unlikely that Marriott's suspicions were derived from Morris; in all probability it was the latter who was put on the track of the imposture by what his friend was able to tell him.

Early in December of 1953 Captain Guy St. Barbe wrote to the Keeper of Geology at the British Museum saying that he—and, it transpired, Marriott also—had reason to believe in 1913 that Charles Dawson had perpetrated a fraud at Piltdown.

In 1912 Captain St. Barbe lived at Coombe Place (near Lewes), taken for a period from the Shiffners. His family entertained much, and at a garden party at Coombe Place he met Charles Dawson and his wife. On one such occasion St. Barbe mentioned his interest in archaeology and showed Dawson a small collection he had made. Dawson asked to borrow some of his Red Crag flint implements. A few months after the announcement of the Piltdown discovery St. Barbe was in Dawson's office, where, after they had finished their professional business, Dawson showed him the Piltdown cast, which Barlow, Woodward's assistant, had just made. This meeting must therefore have taken place in about May 1913. It was a short while later in the summer when St. Barbe went into Dawson's office without knocking. According to Captain St. Barbe, Dawson evinced unmistakable embarrassment. There were some dozen or so porcelain crucibles on Dawson's desk containing brownish-coloured liquid, and there was a strong smell of iodine. Dawson then explained that he was interested in the staining of bones and showed St. Barbe some bones in the fluid. His aim was to find out how bone staining went on 'in Nature', and he said he was trying every possible kind of staining method. A few weeks later St. Barbe was again in Dawson's office. There was no excitement this time, but Dawson intimated that he was staining flints as well as bones.

That was what Marriott, in his turn, saw in Dawson's office. St. Barbe met Marriott socially and was aware of his interest in archaeological matters; he mentioned the staining and so learnt that Marriott had also seen the staining in progress. This was at about the time when the second season's discoveries were being made at Piltdown. (The canine tooth was found on 30 August 1913.) St. Barbe and Marriott discussed and rediscussed it all exhaustively and they became convinced at that time that 'Dawson was salting the mine'. 'That was the expression we used', St. Barbe said. He and Marriott agreed to say nothing for the time being, but to wait till scientific doubt was thrown on the discoveries, which they felt was bound to happen sooner or later.

Their hesitancy is understandable enough. One must realize the enormous weight of authority which buttressed Dawson and his discovery. Famous biologists such as Boyd Dawkins and Sir Ray Lankester (the latter's almost a household name for his popular science writing), had supported Woodward's interpretation; the foremost anatomists, Elliot Smith and Arthur Keith, and leading palaeontologists and geologists such as Newton and Sollas, had all given their massive verdict in Woodward and Dawson's favour. It is not really surprising that the amateurs Marriott and Morris, by some regarded as mere cranks, or that St. Barbe, hardly more than a spectator, and a young man, should hold back from uttering an appalling allegation against Dawson.

In 1914 St. Barbe left Coombe Place to take up his war duties, and he never returned to Uckfield or Lewes. In 1916 Dawson died. In later years St. Barbe told only a few people what he believed had happened at Barkham Manor. Positive evidence that some or all the finds were fraudulent, of course, he did not have. He got to know Keith in connection with an Iron Age skull he discovered in Guy's Rift (named after him), near Bristol,[5] but he said nothing to Keith. In the course of his speleological activities he met Mr. Martin Hinton, who became a close friend. He also knew Kennard, Mr. Hinton's associate.

That A. S. Kennard should have spoken, or joked, about Piltdown as he did will now occasion no surprise. He was in touch with Marriott and St. Barbe, so that their suspicions and Morris's could easily have become known to him. As an old member of the Ightham circle, and a lifelong acquaintance of Lewis Abbott's, he would be well aware of the hopes enter- tained by archaeologists and geologists, local or professional, of finding a late Pliocene deposit in the Weald. He knew in 1926 of the re-survey of the Ouse, which revealed the antiquity of Piltdown to be over-estimated. Could he not have deduced the whole truth, or something like it, that the temptation to invent a 'discovery' of fossil man associated with late Pliocene mam- mals and crude flint tools in a Wealden gravel bed has proved irresistible to someone in or near old Ben Harrison's circle? A. S. Kennard always said he knew who the perpetrator was. But he did not intimate to Mr. Hinton that Dawson was the forger. He died in 1948 and his knowledge of the forger's identity went with him.

Long before the scientific disclosure of November 1953 there was this small interconnected group of persons who had their reasons for maintaining that the Piltdown discoveries were not what the palaeontological world thought them to be. Though Harry Morris, for one, made an outright accusation against Charles Dawson, it is clear that the evidence did not amount to proof of his guilt. Kennard may have realized that despite appearances Dawson could well have been the victim, innocent or coerced, of the real perpetrator.

---

1. Letter from Sir Arthur Keith, 22 November 1953.

2. *Exhibition of Sussex Flint Implements which reveal the Real Culture of the Period of Piltdown Man*, South-Eastern Union of Scientific Societies, 34th Annual Congress, Brighton, June 1929.

3. A remarkable palaeontologist, author of over 250 papers and an authority on molluscs; only in his later years, after retiring from his City post, did he come to work as a 'professional' at the Geological Survey in South Kensington.

4. *Proc. Geol. Assn.*, 1949, **60**, p. 80.

5. Hewer, T. F., 1925, 'Guy's Rift, Slaughterford, Wilts.', *Proc. Spel. Soc., Univ. of Bristol,* **2**, pp. 229–34.

    Buxton, L. H. D., 1925, 'Report on Calvarium from Guy's Rift', ibid, pp. 235–7.

13

The Sussex Wizard

¤ ¤ ¤

It is common knowledge that Dawson did not command high esteem in the archaeological circle of Lewes. Some local archaeologists, on the basis of their personal feelings about Dawson as well as on their long-held, rather low opinion of his archaeological reliability, came to invest the Piltdown discovery with extreme scepticism from the start; objective evidence to back this up, such as St. Barbe, Morris, or Marriott might have offered, there seems to have been none. It is perhaps as well to indicate how Dawson came to acquire his reputation for 'unreliability', since it has a bearing both on the standard of his archaeological work and on the quality of his scientific writing on the Piltdown material, activities which we saw provided grounds for genuine surprise in their vagueness and inaccuracy. The local reasons for Dawson's unpopularity should be assessed as objectively as possible, for it should not be forgotten how solid a reputation Dawson had made with his Wealden collections at the British Museum and how good his standing was with such men as Keith and Woodward.

The deliberate avoidance of the great Piltdown discovery in official local circles is quite undeniable. On my first visit of inquiry in August 1953, I had fully expected to see much made of Piltdown in the local museums. The Borough Museum contained nothing but a small picture of an imaginary *Eoan-thropus* presented by Dr. S. Spokes and some Piltdown eoliths presented by Harry Morris. The Barbican Museum, the home of the Sussex Archaeological Society, hard by the Castle keep, is in the street where once lived the famous Dr. Mantell; and in the same street is Dawson's home, Castle Lodge. Here, too, there were no specimens of Charles Dawson's on view, but more flints of Harry Morris, including eoliths from Piltdown. A cast of the well-known reconstruction of Piltdown man was displayed along with three enlarged models of teeth—one the molar from Piltdown II and, for comparison, chimpanzee and human molars. The cast and models (and also the picture in the Borough Museum) had all been presented by Dr. S. Spokes in 1928—fifteen years after the world-famous discovery. Nothing else of the Piltdown assemblage was on view.

I wondered whether any record of presentations from Dawson (or his estate) would be found in the *Sussex Archaeo-logical Collections*. I looked through these volumes for some detailed account of the Piltdown discoveries, for information such as the names of those interested, discussion at local meet-ings, accounts by eyewitnesses of the stirring events, headline news in those days, and hoping in these records to come across some telling discrepancy or other clue. I was not at all prepared to discover on reading through these volumes of the years 1911 to 1916—no mention of Piltdown at all! No meeting had ever been held, no address given by Dawson, by then their most famous member. Only in 1925 did the Sussex Archaeological Society hear a talk on *Eoanthropus dawsoni*, from Sir Arthur Smith Woodward, who in his retirement was now living nearby at Hayward's Heath. The text of this paper was not published.

This indifference to or rather disregard of Dawson's discovery recalled to mind the long delay which had elapsed before the Society exhibited the cast in its museum. Later I learnt that I was not the first to be astonished at the local neglect of the Piltdown find. Mr. L. V. Grinsell had the same experience in 1927 or 1928. Intending at that time to compile a book on Sussex archaeology (an aim abandoned when Cecil Curwen's book came out in 1929), one of his first tasks was to work through the collections, and he too was struck by the lack of reference to Piltdown. So strange did it seem to Mr. Grinsell that he mentioned his surprise to the late Dr. Eliot Curwen. I, too, had noticed that Dawson had never served on the Council in all his twenty-five years of membership, that there was no mention of his death in 1916, and no obituary notice. Nor was the Society represented at his funeral. All this took on the look of a deliberate avoidance of the whole Piltdown affair and of Dawson personally.

Still, the archaeological collections contained (besides Dawson's own communications on iron-work and other topics), two references to Dawson's work, and though unconnected with the events at Barkham Manor, they are of some relevance.

A note in 1909 from Mr. Ray,[1] local Secretary of the Hastings Branch of the Society, records the occasion when Mr. Dawson accompanied him on an excursion early in February 1907, to view two Neolithic skeletons, just unearthed on the Duke of Devonshire's estate near Eastbourne. Dawson came into this by chance, for Ray was to be accompanied by Mr. Salzman, one of the committee at Lewes, but the latter had to go up to London. This investigation has already been mentioned for the apparent care, according to Ray's report, with which the remains were examined, for their positions were carefully recorded, measurements made, and photographs taken—all in obvious contrast to the conduct of those operations to be undertaken at Barkham Manor in the not too distant future. We are inclined to think that the careful observer was Mr. Ray, a belief justified by the other reference in the *Collections* which throws an unkind light on Dawson as an archaeologist.

The second paper is by the then well-known authority, Hadrian Allcroft,[2] and was read in 1916; it surveys, in a thorough manner, some earthworks of Sussex and deals at length with Dawson's work at the Lavant caves. Allcroft is outspokenly critical about Dawson's investigation. He deplores the poor field work,[3] 'the skill of a North Country miner would have dealt easily with the matter at the outset. . . . As it is the caves, it is to be feared, are now lost for all time and their secrets with them.' There was to have been an extension of the work but this Dawson had failed to carry out; and 'to make matters worse he had never produced the promised paper', though he had reported to the Chichester Society and also given an interview to the local paper in 1893; the finds as far as they were known were complex and Allcroft's irritation is easily understood.

The actual finds made by Dawson and Lewis are certainly unusual, and Allcroft found some difficulty in arriving at a satisfactory interpretation. There were objects denoting Neolithic flint-mining, evidences of Roman earthworks, and medieval wool seals. Cecil Curwen comments on the strangeness of this collection, but refers the reader to an interpretation given by his father, Eliot Curwen,[4] which in fact accords with certain of Allcroft's views. Apart perhaps from the unsatisfactory prosecution of the work there seems nothing in this rather interesting and probably unique discovery to bring actual discredit to Dawson. Yet it certainly seemed to have left a number of archaeologists extremely distrustful,[5] a view of Dawson known to have been shared even by Eliot Curwen. As Dawson and Lewis did their work in 1893 and Allcroft made his forcible comments only in 1915, we can understand that dissatisfaction over the outcome must have existed, to Dawson's growing detriment, during all that period. During that time, too, Dawson became involved in the affairs of the Sussex Archaeological Society in a manner which earned him lasting personal unpopularity.

To follow up what I had learnt from the records of the Sussex Society, I got in touch with Mr. L. F. Salzman, one of the few survivors from those early years and still as active as ever as Hon. Editor of the Society's publications and a member of its Council.[6]

In discussion with him on that first day of my inquiry, I expressed my surprise at the apparent neglect of Dawson and his discovery by the local Society. He agreed that this was indeed the case, and said that Dawson's activities had come to be received sceptically, partly on account on his archaeological work (this I took to refer to the Lavant Caves), partly on account of his historical work on Hastings Castle, which had been not well received locally, but largely because of the 'Castle Lodge' episode.

Castle Lodge, then owned by the Marquess of Abergavenny, was used by the Sussex Archaeological Society as a meeting-place and museum. In the autumn of 1903 the Council received an intimation that the Lodge, occupied by the Society since 1885, had been bought by Dawson, who soon afterwards served upon the Secretary formal notice that the Society was to terminate its occupation of the premises at midsummer 1904. This purchase by one of its own members took the Council by surprise, as its official record of the matter states,[7] since it had been understood that if the property were to be offered for sale the Society would have the option of acquiring it.

The Society was not alone in its consternation over Dawson's purchase, which caused them much inconvenience for several years. The vendors, too, were taken aback, for they had not realized until the last stages of the sale what Dawson had been about. This Mr. E. V. Clark of Lewes (a man well disposed to Dawson) heard from his close friend, Mr. Arthur Huggins, the Marquess's Agent.

It is perhaps not so surprising that in 1907 when he moved from Uckfield into the recently acquired Castle Lodge, which meant encountering the daily coolness of the recently evicted tenants (who found new premises a few yards down the road), Dawson now joined the Hastings Natural History Society and was seen hardly ever at meetings in Lewes. It must have been altogether a socially trying environment for Charles Dawson and his wife (he had married Mrs. Helène Postlethwaite in 1905), a charming and cultured woman, there in Lewes, with its still Victorian outlook (as a friend recalls) and the unfriendliness and indifference of the local archaeologists. Those ten years, the last left to them both, the years of Dawson's fame, ended sadly for Mrs. Dawson. Seriously ill herself at the end of Dawson's prolonged illness (he was nursed devotedly by his step-daughter), things were very difficult when he died. The widow sold the loan collection to Hastings Museum through their friend, Mr. W. R. Butterfield, and Sir

Arthur Smith Woodward helped her in the matter of a Civil List pension. She did not long survive, and Castle Lodge was sold in 1917.

Had it not been for this Castle Lodge incident, would Dawson, however ambitious or 'proud' he may have been (as he was according to some of his local colleagues), have earned hostility merely as a result of his Lavant work, or of the weaknesses of his imposing two-volume work, the *History of Hastings Castle?* Though this has become a standard and useful work of reference, it was early recognized as less a product of genuine scholarship than of extensive plagiarism. Of this work a contemporary reviewer in the *Sussex Archaeological Collections*,[8] says that it contains a great deal of material relating to East Sussex, mingled with a certain amount of general history, 'not always accurate', and it goes on:

> The author has displayed much industry in collecting material, but little judgement in its selection and arrangement. Apart from errors of translation, the misreadings are extremely numerous. It is difficult to say how far these are due to carelessness, inaccuracy and neglect of proofreading and how far reliance on second-hand authorities, as references are frequently omitted or given in unintelligible form. In many cases where matter is taken, mistakes and all, from earlier writings, no acknowledgement of the source is made.

The reviewer's conclusion that Dawson had claimed far too much for himself in the work is a view which Mr. Manwaring Baines, the present Curator of the Hastings Museum, fully endorses. A year before the Piltdown exposure of 1953, Mr. Manwaring Baines, who was himself studying the history of Hastings Castle, came at last into possession of the manuscript of William Herbert, the antiquarian who carried out the 1824 excavations. This is the work to which Dawson makes only scanty acknowledgements. Mr. Baines[9] declares that half the material in Dawson's volumes is copied unblushingly from Herbert's manuscript, and describes the rest as gross padding.

Yet, when all is said and done, it seems certain that the Castle Lodge incident did much, if not everything, to secure for the versatile, ever-inquiring, and energetic Dawson a bad reception locally for his *History*, while his untidy cave exploration made things worse. When his great discovery came at Piltdown, it elicited profound scepticism.

Yet the Castle Lodge episode distorts and obscures the real character of Dawson, the collector, geologist, archaeologist, and antiquarian, of whom Woodward wrote:[10]

> He had a restless mind, ever alert to note anything unusual; and he was never satisfied until he had exhausted all means to solve and understand any problem which presented itself. He was a delightful colleague in scientific research, always cheerful, hopeful, and overflowing with enthusiasm;

and Keith[11] admired him unreservedly. He noted in his journal (28 January 1913): 'Charles Dawson comes to see me. A clever, level-headed man.'

The obituary notices of Dawson—his death on 10 August 1916 was widely noticed—all testify to the remarkable range of his interests and the striking nature of many of his finds. As a collector, Dawson was most effective in the palaeontological field with his Wealden specimens, but his avidity found expression in his assemblage of iron-work and other archaeological objects and etchings and paintings of antiquarian interest. The loan collection bought by the Hastings Museum makes a remarkable list—it includes items of iron, bronze, stone, glass, jade, and bone. His enthusiasm extended to many varied additions and alterations to the fabric of Castle Lodge: in the windows, doors, gates, and banisters there are pieces obtained from medieval and other old buildings. He also took an interest in antique furniture and photography. He drew clearly and accurately.

Piltdown man can fairly be viewed as the climax to a whole series of out-of-the-ordinary discoveries extending over thirty years: some, like his Wealden mammals, of first-rate scientific value; some spectacular, like the discovery of natural gas at Heathfield; and not a few quite unprecedented, like the Roman figurine from Beauport Park,[12] the Ancient Boat at Bexhill,[12] the transitional horseshoe,[12] and the bone implement from Piltdown; and some, again, quite bizarre, like the 'Toad in the Hole' in the Brighton Museum. It is not surprising to learn from the obituaries that Dawson had come to be known, not inappropriately, as 'The Wizard of Sussex'.

These many pursuits, so remarkable in scope and variety for an amateur who was at the same time an extremely busy professional man, what do they tell us of his quality as an investigator? With Keith and Woodward and the Geological Society, Dawson's prestige as geologist and palaeontologist stood high, but locally, as Sussex archaeologist and historian, he was far from being so well regarded. In London, toward the end of 1913, some of his geologist colleagues proposed to send his name forward for consideration by the Royal Society; in Sussex, at that very time, there were those who privately accused him of the grossest scientific malpractice. Personal feelings about him were in keeping with these extremes of judgment on his scientific worth. Keith, Butterfield, and Willett, like Clark or the Kenwards, liked and admired him, and found him easy to get on with; but, as we know, there were others with very different views, and it appears that after years of close collaboration there was a serious quarrel and break with John Lewis about 1911. What picture of the man emerges from a more detailed scrutiny of his diverse undertakings, the fruits of his 'restless curiosity' and 'sharpness of sight that never missed anything of importance'? Which of the opposing evaluations of the man do they support, and can they help to throw light on the evidence we have adduced which show Dawson to be in some way peculiarly entangled in the dark enterprises of the Piltdown affair?

The spectacular as well as the original element in the man is exhibited in nearly all his activities. It is illustrated, for example, in his announcement of the discovery of natural gas at Heathfield, used for many years for lighting the local railway station; at the reading of his paper to the Geological Society in 1898, the lecture-room that night was illuminated with gas brought specially from Sussex. The occasion was somewhat marred, for in the discussion his chemical understanding of the nature of the gas was called seriously in question. There was the discovery which Dawson made in the Museum of the Royal College of Surgeons. He wrote in high excitement in May 1912 to his friend at the British Museum to tell of 'a thirteenth thoracic vertebra' he had lighted on in the skeleton of an Eskimo. Dawson never wrote this up, but the newspapers (I am told) 'ran' it during or after the Piltdown days as the discovery of a new race by the discoverer of Piltdown.

Another example of Dawson's 'restless curiosity' was an investigation of the development of an 'incipient horn' in a cart-horse, an abnormality which attracted his attention during the beginning of his last illness.

As an oddity, this reminds one of the 'Toad in the Hole' which he presented to the Brighton Museum in 1901 through Mr. Henry Willett. This is a petrified toad in what is actually a hollow nodule of flint and is still in the museum. No one has claimed that this is a true fossil; the toad when young must have got into the nodule through a small hole and found enough insects to enable it to grow until it became too large to get out again. The specimen was actually found by two Lewes workmen, Thomas Nye and Joseph Isted, in the summer of 1898, and was examined on the day of discovery by Dr. J. Burbridge of Lewes, who testified to the integrity of Nye. How it came into Dawson's possession has not been ascertained, but it was exhibited by him at the Linnean Society on 18 April 1901, and figures in the *Illustrated London News* and *Graphic*.[13]

In the years before he brought Piltdown to Woodward's notice, Dawson was constantly calling his friend's attention to all manner of new and strange things. There were the palae-ontological finds, dinosaurian, mammalian, botanical and insect, and some even more curious examples of animal life—a cross between a carp and a goldfish in 1909 and, with great circumstantial detail, a 'sea-serpent' observed in the Channel on Good Friday of 1906. So varied were Dawson's ventures and so inexhaustible his energy that in 1915 we find accounts in the newspapers of his experiments with 'flaming' bullets—phosphorescent anti-Zeppelin bullets. This is not really a sur-prising departure. The necessity of such a missile was urgent enough and the experiments may reflect some connection with Metro-Vickers and Maxims (of which his brother was Managing Director).[14] It accords with our estimate of Dawson's originality, ingenuity, and flexibility of mind.

As with palaeontology, in iron-work Dawson took a

sustained and serious interest and achieved something of a reputation as an authority on that subject. He made his own extensive collection of iron objects, and some of these were shown in the Exhibition of Sussex Iron-work and Pottery he organized for the Society at Lewes in 1903, and again at an exhibition at Hastings in 1909. For the former he wrote an article of twenty-seven pages on iron and five on pottery,[15] and for the latter he provided some notes in the Catalogue. What do his writings and his collections reveal?

His writings on iron are much like his *History of Hastings Castle*: they are useful compilations which must have cost him much labour, and, like the *History*, they are not authoritative. The greater part has been taken from an early writer, Topley, almost word for word without acknowledgement; the 1909 article is full of errors many taken over uncritically from earlier authors.[16]

The most interesting iron object in Dawson's collection is the statuette from Beauport Park which he bought, some ten years before exhibiting it, from a workman who found it in 1877 along with coins of Hadrian's time. It has been the cause of considerable controversy since its first showing at the Antiquaries in 1893 by C. H. Read, who pointed out that if it could be proved to be Roman, as Dawson claimed from its reported site of discovery, 'the discovery would be one of great importance, seeing that Roman works of art in the round in iron are of the highest rarity'. Opinion at the meeting was decidedly against the claims of its Roman character. Straker[17] attributes its uniqueness to its spurious character and Mr. Downes concludes that the statuette was nothing more than a nineteenth-century product which was sold as a curio. The matter cannot be conclusively settled, but the evidence is against Dawson's view, though he was entirely correct in judging the figure to be of cast-iron.[18] From Rock's description of the Beauport cinder heap before it was cleared away, Mr. Downes considers it most improbable that Romano-British furnaces could have produced cast-iron. He points out how odd it is that James Rock in his article of 1879 should mention a coin of Hadrian but not the statuette supposed to have been found with it.

As with the Lavant Cave finds, or those at Piltdown, or the 'transitional horseshoe',[19] Dawson's documentation all too frequently turns out to be disquietingly faulty or vague.

In his last Piltdown year of 1915 Dawson embarked on a variety of activities. In February he read his paper on the vexed and burning question of 'eoliths' to the Royal Anthropological Institute. This again illustrates his ingenuity, and is of some interest, as it reveals his relations to Lewis Abbott in the last years. Dawson took a definite stand in opposition to such workers as Reid Moir (whose star was in the ascendant), Abbott, and others. He brought forward a telling and novel demonstration, using starch models, of the ease with which mere natural agencies would produce the prismatic fractures of flint. By shaking or sitting on a bag of these pieces of starch, Dawson reproduced all the well-known 'eolith' shapes. The exhibits are still to be seen in the British Museum, and Dawson made his specimens of starch all the more convincing by staining them the colour of natural flints. The *Lancet* of 13 March 1915 noticed his paper, but Dawson found the difficulties of the war years too harassing, as he wrote to Woodward,[20] to allow him to prepare a report for publication. Dawson suffered some invective from Abbott as a result of his views on 'eoliths'.[21] In his previous utterances on the subject (in 1913) he had been circumspect in his remarks, as we know, but he was now obviously on stronger ground with his experimental models.

He was prepared to stand up to and criticize Abbott severely on another matter about which he wrote on two occasions to Woodward, in April and May of that same year. Abbott at the time was making much of new 'Pliocene' strata he claimed to have discovered in Hastings. Dawson went to look for himself and found himself in violent disagreement.

> I do not know [he wrote] what the Geological Association will say as to Lewis Abbott's Pliocene river beds and deposits . . . I can only see beach stones from the shore mixed with surface soil from farmyards and road scrapings, put on the land for manure. A small line of beach flints 3 inches thick adjoining an old hollow road and shown in the section of the bank is the remains of an old road mending when the road ran at a higher level some years ago. J.W.L.A. calls this a Pliocene river bed!

And in the other letter he reiterates his views, adding: 'We shall soon hear of pre-historic balls discovered at Ipswich and laid at the feet of the East Anglian Scientific Society.'[22]

Dawson's acumen as a geologist appears again in this matter. When Mr. Edmunds[23] in 1926 made the important visit to Hastings, Abbott pressed on him his claims to have found a Hastings Lower Pleistocene bed and Mr. Edmunds, in his turn, was greatly astonished at this section. In 1915 Dawson had confidence neither in Abbott's eoliths nor Pliocene beds, a striking contrast to the deference shown to Abbott's views on these subjects three years before. Clearly, 'the colleague on the Museum Committee' was not so dependent on Lewis Abbott's counsel as might be thought from the consultations which Dawson had with him in the first years on the Piltdown cranial finds and the palaeoliths and from Abbott's article on the Piltdown discovery.

The exuberance of Dawson's enterprise should not make us think of him as a dilettante, a mere collector of the odd and curious. His central interests—his palaeontology, his iron-work, and his *History*—all were pursued with perseverance and an undeniable perspicacity. In his work there is the mark of great nimbleness of mind, of imaginative insight, and of the flair for a discovery just such as Piltdown in its setting might so easily have been.

There is a consistent pattern running through all Dawson's endeavours. It is not merely the novel or spectacular element which is characteristic so much as the persistence of his con-cern for the 'transitional' or, to use his favourite term, the 'intermediate' form. Such in fact were the things which brought him the notice of palaeontologists and archaeologists alike. His first notable discovery, like his last, was of a 'missing link'. Did *Plagiaulax*, the mammal with its reminiscence of the reptile, fire him with the notion that evolutionary 'links' were the supreme prizes? In the long period between *Plagiaulax* and Piltdown, he brought forward such examples as a 'transitional' boat, half coracle and half canoe (1894),[24] a 'transitional' horseshoe (1903),[25] a neolithic stone weapon with a wooden haft (1894),[26] the first use of cast-iron (1893), a form between *Ptychodus* and *Hybocladus* (1903),[27] a Norman 'prick spur',[28] a cross between goldfish and carp (1909).[29] That Dawson should thoroughly appreciate the full implications of a 'Dawn man'— faunistic, geological, archaeological, and even anatomical— emerges irresistibly from the record of his activities, his abilities, and his habit of thought.

With all his gifts of drive and imagination, and though his

knowledge in his chosen fields was sufficient to enable him to meet on their own ground those professional geologists, zoologists, archaeologists, and even anatomists whose attention he roused with his discoveries, Dawson yet showed himself ill fitted for the exacting work of accurate historical research or of well-documented field investigation.

There emerges also from our survey the certainty that Dawson was always eager for new and arresting discoveries. His anxiety for recognition is clearly displayed over his belief that at the Royal College of Surgeons he had found a new race of man. He writes to Woodward on 12 May 1912 in the midst of the Piltdown excitement:

> Since I saw you I have been writing on the subject of 'the 13th dorsal vertebra' in certain human skeletons which I believe is a new subject. I send you the result and if you think well enough of it I should be very much obliged if you would introduce the paper for me at the Royal Society. I am very anxious to get it placed at once because I have had to work the photographs under the nose of Keith and his assistant.
>
> I gather from the latter person that Keith is rather prickled as to what to make of it all, and I want to secure the priority to which I am entitled.

In his time Dawson's scientific successes had been somewhat uneven. When he and Woodward began their scientific careers, signified by their simultaneous election to the Geological Society in 1885, the amateur, in his spare time, under the tuition of old Mr. Beckles, had already made his mark. *Plagiaulax* in 1891 (and 1911) added greatly to his reputation, and the iron-work exhibition of 1903 was an important local event. But the *History of Hastings Castle* of 1909 proved something of a disappointment, and there were other setbacks (such as Lavant and the statuette) and the unpleasantness of the

Castle Lodge incident. When he renewed his scientific contact with Woodward, Dawson wrote in March 1909 of 'waiting for the big discovery which never seems to come'. Piltdown brought him public acclaim in high measure, though scientific recognition was slower. In February 1914, in a letter to Woodward congratulating the latter on his election to the Presidency of the Geological Society, Dawson voiced his sense of neglect: 'I have a feeling that the Geological Society are treating me rather shabbily'. His death at the age of fifty-two in 1916 cut everything short.

Dawson left no information for Smith Woodward. He told his wife little of his Piltdown doings. He was extremely secretive about his excursions, as Mrs. Dawson told a friend and lamented again after his death, for she felt that at the end he had had some important information he wanted to pass on. We know of Dawson's secrecy also from the recollections of the late Mr. E. V. Clark and his wife, for when they dined at Castle Lodge in the autumn of 1911 (or perhaps early 1912) Mrs. Dawson remarked playfully about the 'secrets of the cellar'. Dawson's own papers, as far as can be traced, were all destroyed—those at Castle Lodge in 1917, and those at Uckfield (which included the manuscript of the *History of Hastings Castle*) were sent for salvage in World War II.

One 'document' only was left—the Barcombe Mills skull, meriting its title of a 'third Piltdown man' only in the artificial treatment it has received. With this 'document' as our last piece of evidence we must attempt finally to arrive at an understanding of Dawson's part in the fossil fraud.

---

1. *Sussex Archaeol. Collections*, 1909, **52**, p. 120.
2. *Sussex Archaeol. Collections*, 1916, **58**, p. 65.

3. Ibid., p. 74.

4. *Sussex Notes and Queries.*

5. Mr. Salzman in *Sussex Express and County Herald,* 27 November 1953. See remarks by Mr. A. W. Oke, quoted earlier.

6. Elected President of the Sussex Archaeological Society in March 1954 in succession to the Duke of Richmond and Gordon.

7. *Sussex Archaeol. Collections,* 1904, **47**, p. xiv.

8. *Sussex Archaeol. Collections,* 1910, **53**, p. 282.

9. Letter, 19 February 1954. Mr. R. L. Downes has made a detailed study of the book and is in complete agreement with Mr. Baines.

10. *Geol. Magazine,* 1916, **3**, pp. 477–9.

11. Personal communication, 22 November 1953.

12. These objects are described later.

13. For this information I am indebted to Mr. C. Musgrave, F.L.A., F.M.A., Director of the Art Gallery and Museum, Brighton.

14. It was this brother, Sir Trevor Dawson, R.N., who presented Charles Dawson to King Edward VII at a levee in 1906.

15. *Sussex Archaeol. Collections,* 1903, **46**, pp. 1–62.

16. I am indebted to Mr. Downes for this information.

17. Straker, E., 1931, *Wealden Iron,* pp. 335–7, London.

18. Report by Mr. Morrogh of the British Cast Iron Research Association.

19. *Sussex Archaeol. Collections,* 1903, **46**, pp. 23–4. Dawson had in his collection an iron horseshoe which he claimed incorporated the features of both the Roman 'hippo-sandal' and the modern horseshoe: this slipper-like specimen could be strapped on and had holes for nailing on to the hoof.

20. 8 October 1915.

21. Letter to Woodward, 9 March 1915.

22. Interestingly enough, the bolas stone is now recognized as an important tool of the Chelles-Acheul culture. At Olorgesailie in Kenya Leakey found some good examples of thee spherical stone balls with the Handaxe culture.

23. Personal communication.

24. *Sussex Archaeol. Collection,* 1894, **39**, p. 161.

25. ibid., 1903, **46**, pp. 23–4.

26. ibid., 1894, **39**, p. 97.

27. Letter to Woodward, 3 November 1903.

28. Exhibited, Hastings, 1909: Mr. Downes informs me that the object is not a prick spur and cannot be identified.

29. Letter to Woodward, 13 July 1909.

14

The Question of Complicity

¤ ¤ ¤

Thus minded canst thou safely venture.
Resolve thee! Set thine hand unto the indenture!
With joy mine arts forthwith thou'lt see.
What no man yet beheld, that give I thee.
                    GOETHE: *Faust*, Part II.

In 1908 (as far as we can ascertain), long before he went to see his 'old friend' the Keeper of Geology at South Kensington, some time before he met Teilhard de Chardin, Dawson had in his hands the first piece of the skull of *Eoanthropus*. This piece, we know, had been chemically treated, by iron sulphate, to produce the brown colour, and in the process the bone had undergone the change in its crystal structure revealed by the X-ray diffraction method. This piece was part of the brain-case, the 'coconut', smashed by the labourers, according to the story the origins of which are by no means clear. Once again a question faces us which raises sharply and finally the issue of Dawson's complicity. We might put the question as the 'Piltdown Riddle':—Was the pit completely barren at the birth of Piltdown Man or did he begin life there as a burial? Was the cranium genuinely in the gravel or had it been planted where the workmen found it?

In the first stages of the investigation, before we fully appreciated the artificiality of the iron-staining, we were inclined to regard the skull-case in the gravel as a genuine though not very ancient fossil. The fluorine values, while not really high, taken with the reduced content of organic matter, certainly gave grounds for accepting a semi-fossilized condition in the cranium. So it was presumed at first that the hoax had been based on a genuine discovery of portions of an ancient skull in the gravel, and that the ape jaw and canine and the other animal remains and implements had been subsequently planted. As the investigations went on, stage by stage, this view became untenable. The iron-staining threw serious doubt on the skull's derivation from the gravel; the sulphate in the bone, in the form of gypsum, is the result of artificial and deliberate chemical treatment, and gypsum does not occur in the Piltdown or Barcombe Mills gravel. The chemical conditions in the Piltdown subsoil and gravel water are not at all such that this unusual alteration in the bone could have taken place naturally in the gravel. On the other hand, the cranium, if genuinely discovered by the workmen, may have tempted the Piltdown forger, because of its mineralized appearance, its provenance in a likely gravel bed, and its unusually thick walls, to enhance its palaeontological value by removal of various pieces of bone and to establish its antiquity by improving on its colour.

To suppose that the 'coconut' really came as an authentic local fossil from the gravel or the loam is to incriminate Dawson completely, since the staining would necessarily have taken place while the cranium was in his possession and with his knowledge. The evidence does not allow of this quick answer to our riddle. We do not believe that the cranium was a genuine local fossil. We believe it to be intrusive exactly like all the other 'finds', animal and archaeological (excepting the eoliths). The reasons for this belief are:

(1) The gravels outside the Piltdown 'pocket' seem entirely unfossiliferous even of small animal bones (except perhaps for a few surface modern specimens), and the chemical conditions are highly unfavourable for the preservation of bone.

(2) The fluorine value of the brain-case, though higher than that of the jaw or of modern bones, appears still far too low for the geological age of the terrace, which may in fact be as early as the end of the Middle Pleistocene.

(3) The hippo remains, which have the same fluorine value as the calvarium and on which reliance was placed as a reasonable 'marker' for an Upper Pleistocene date for the brain-case, turned out to be intrusive, for their iron- and chromium-staining is not the work of Nature.

(4) The fact that the cranium, too, has been deliberately iron-stained is on the whole against its *in situ* provenance for even a late prehistoric burial here would very likely have been iron impregnated and would not have required treatment, since the ground water here is very rich in iron.

(5) Our scrunity of the accounts of the digging up of the various fragments has given us no confidence that anything did come from undisturbed soil, despite Dawson's assertions.

These are weighty reasons against the authenticity of the cranium as a local fossil. Yet we cannot easily dismiss the story of the gravel-diggers and their 'coconut' as pure invention, a plausible tale put about to furnish an acceptable history for the pieces of 'spoil' which were being shown round and talked of in 1911 or before. The contemporary documentary records are indeed very unsatisfactory, as has already been pointed out, and for the acquisition of the first cranial piece we have only Dawson's own vague account of 1912 to go on, and for an explicit mention of the 'coconut' only that by Abbott and another in the *Antiquary* in February 1913. It is hard to believe that this story might have been a later invention, seeing that Dawson gave out the first news to Woodhead very probably in 1908, and the two friends could hardly have gone back for their immediate search without the knowledge of the tenant, Mr. Robert Kenward. Miss Mabel Kenward feels sure her father was apprised of events at that time, though whether he was first informed of the find by the labourers remains uncertain. Dawson himself heard of it from the labourers[1] who had also kept a piece to give to him (though it is possible that they may have been acting under instructions from Mr. Kenward). Dawson told frequently of the labourers' part (even if he did not clearly record the coconut episode) in the next few years and could hardly have had reason to fear anyone's seeking confirmation of the men.

Granting, then, probability that the workmen did find a portion of skull, it is still conceivable that what they found was not the semi-fossil *Eoanthropus* but some very recent and quite ordinary burial. This uninteresting skull could have been quickly retrieved in that shallow pit, and thereafter a substitution made, in stages, of the much more arresting fossilized, suitably treated, thick skull which happened to be in the culprit's possession and which (with the stimulus of the cranium-less Heidelberg jaw) may well have started him off on the enterprise. Even if some pieces of the 'original' skull were overlooked they would almost certainly have disappeared in the next three or four years as the gravel was removed for roadmaking—a fate which Dawson[2] thought must have overtaken some of the 'real' Piltdown fragments. Moreover, there is little or no guarantee in Dawson's own records that the excavations of 1912 and later were made on the spot where the labourers first found their piece of bone—a piece which in fact had been in their possession for some unspecified time. There were no landmarks set and the area was frequently under water.

At first sight the nasal bones and the turbinal uncovered in the upper disturbed[3] layer of the gravel seem the probable remnants of just such a recent burial, for these structures would otherwise hardly have survived. But the turbinal has been proved to be no turbinal at all,[4] and though the nasal bones belong on chemical grounds to a skull other than *Eoanthropus* the bones are intrusive—they carry the sulphate and gypsum.

If this suggestion of the substitution of the famous 'Man' for an indigenous skull could be sustained—and it is inherently possible—Dawson would again be inescapably incriminated. It would occasion no surprise if a collector so versatile should have acquired human skulls, usual or unusual. He had come across them in previous excavations (e.g. 1906) and he had the Barcombe Mills specimens when he died.[5] Doubt has in fact been thrown on his assertion that there was nothing to be found at the site except the one fragment after the labourers had dispersed the coconut. This challenging information merits the most serious attention for it comes from the informant on whom we rely for the only independent confirmation of the 1908 date, but we must admit that the evidence is not conclusive.[6]

On balance we should assume that the gravel started as an entirely 'sterile' deposit, that in it a largish portion of the brain-case of *Eoanthropus*—to be—was 'planted', and this the road-menders broke up, retaining just the one piece. Some of the pieces which were not irrevocably lost in the road metal may perhaps have been retrieved later.

Who then was the author of the 'plant' and the plan, though it took four years to develop, for foisting a 'pre-Heidelberg man' complete with fauna and implements on to the receptive palaeontological world of 1908? This plan, hatched in 1908, as far as can be ascertained, by someone well aware of the gravel deposit and the gravel-digging, was put into operation before Woodward or Teilhard had even heard of the site. Does the evidence allow Dawson's role, like that of Woodward and Teilhard, to be viewed as that of a mere dupe?

If Dawson was an innocent victim, a dupe from start to finish, we should have to accept a reconstruction of his behaviour which would include a number of very remarkable episodes, for in many ways, as we have learnt, his actions coincide so surprisingly with those of the perpetrator.

Dawson's admission to Woodward of his use of bichromate seems, on the face of it, the action of one oblivious to the widespread implications of this information. Yet, as we have seen, we cannot in fact be sure that Woodward did hear of it from his co-worker. Be that as it may, this use of bichromate in the winter of 1911–12 proclaims an identity of practice in just those respects necessary to further the fraud. Of the miscellaneous remains brought to Woodward, the perpetrator used chromium on all except the five cranial pieces, for these

Dawson had by some coincidence already so treated. And, again, while Dawson dealt with the occipital, parietal, and temporal bones of Piltdown I in this way, the unknown perpetrator did just the same to the frontal bone of the same skull which came to light in 1915. If Dawson's activities were innocent, they turned out at the same time to be quite complementary to those of the culprit. The iron-staining emphasizes this strange conformity. Where Dawson had super-added the chromate 'mistakenly for hardening', the perpetrator had done this deliberately as part of the staining of fossils which needed no hardening—for example, 'Stegodon', hippo, and flint E.606.

Could Dawson have been oblivious of the iron-treatment? The Barcombe Mills 'skull' in his possession had adhering to it a piece of matrix which clearly reveals the iron-staining process which was used to produce the 'gypsum'. We have here a specimen which could hardly have been foisted on him as a 'third' *Eoanthropus*, for he made no special claim for it and left little or no information about it. A straightforward anatomical examination would show that it had little in common with Piltdown I, and was in any case a composite affair of several pieces of different skulls. The staining experiments witnessed by Captain St. Barbe and Marriott were on a fair scale: Dawson explained that he was trying all sorts of chemicals to see how staining occurred in Nature. What a coincidence, then, that he should possess this all but incriminating Barcombe Mills specimen. As we can see no basis for claiming this as another fossil in the series, we find it difficult to avoid the suspicion that the 'gypsum' staining was done by Dawson himself, presumably as one of his experiments.

On the view that Dawson remained completely deceived,

and if we accept without scepticism the whole story of the different discoveries, we are faced with another set of curious circumstances. At site II at Sheffield Park amazing luck attended Dawson's search. In less than half a year he found two quite small pieces of bone and one tooth in the raked-up stones of a ploughed up field; he found these *seriatim* and they established the fact, not only of a second individual, but a second Piltdown man. All this contrasts with his ill-luck at site I. Here in a small and restricted gravel pit it took him some three or four years to find the second rather large piece of skull from the 'coconut' which was known to have been smashed into several fragments, and this though he was soon on the spot and there were eyewitnesses of the scattering of the broken pieces. Nevertheless, by the time he was ready to see Woodward he had accumulated an impressive assemblage; how it happened we do not really know. Dawson's activities at both sites seem to serve the purposes and timing of the perpetrator uncannily well.

If these finds at site II were genuine discoveries, why should Dawson fail to give Woodward precise details? Woodward may have been at fault in not securing the information before Dawson's illness made his inquiries fruitless, but Dawson himself had had plenty of opportunity of providing the necessary information before he fell ill. Are we then to regard all Dawson's omissions of recording and his various inaccuracies as merely glaring examples of the deficiencies in his archaeological and historiographical abilities, as evidenced in some of his other work? The difficulty here is that these deficiencies in the Piltdown record once again serve the purposes of the fraud, for they shroud the early history in obscurity, they invite the erroneous belief that much more was found in Woodward's time than was the case, and the ordinary reader might conclude that the jaw and the cranium (because of the occipital fragment) were both found *in situ* in practically the same spot.

Let us turn, however, to the alternative possibility and see whether those same queer circumstances may not be more satisfactorily understood on the view that Dawson was indeed not oblivious to the true state of affairs, but had participated in them progressively as victim, detector, and unwilling accessory.

On this view, his knowledge of the bichromate treatment may have derived from his knowledge of its use in the hoax, and he would presumably have come on this revelation rather late in the affair. If this is so, it is hard to see why Dawson should have told Woodward about it at all, and that would only go to strengthen the possibility that Woodward never heard of it from Dawson. We can hardly suppose that the latter learnt of it at an early stage—say, before the public announcement of December 1912—and was yet content to go on with the hoax with his eyes open! If his discovery of it could only have been a late one, we are back at the same point: that Dawson had admitted doing, apparently by sheer coincidence, the same kind of thing as the perpetrator, and had therefore helped in the fraud inadvertently. Whichever way we look at this chromium-staining, it is not susceptible to any explanation which clearly dissociates Dawson's actions from the perpetrator's.

If the chromium-staining remains a serious matter, the Barcombe Mills fragments on the present theory do obtain a possible explanation as the product of his experimental detection of the hoax, and the staining witnessed by St. Barbe and Marriott appears as an episode in his detection of the hoax. But this was still in 1913, and Dawson seems therefore to have fallen in with the hoaxes of 1914, and 1915.

We can argue that it was only Dawson's unfortunate predicament which led him to produce the four pieces at site II in so curious a manner from so unlikely a place, and to evince an understandably small inclination to go into details with Woodward. It implies that here, too, Dawson knowingly perpetrated the forger's plan. He would therefore have been in active, if unwilling collusion, in the matter of Piltdown II, at the very least. A desire to make the most of the discovery could be evoked to explain the unsatisfactory features of the account of such finds as the bone implement, the occipital fragment, and the nasal and turbinal bones.

As in the case of the chromium-staining, there are still features which this explanation of victimization cannot clear up at all convincingly. Why should Dawson at the very beginning provide vague and confusing accounts of the earliest stages of the discovery, which without Woodward's last book and the letters make it impossible to discover what in fact he found before 1912?

The theory of Dawson's complete innocence or obliviousness leaves every sinister event unexplained; the theory of Dawson's late detection of his own victimization does him no credit and still leaves much that is disquieting. The second theory requires that Dawson suffered coercion, but there is no trail, except conjecture, leading to a possible blackmailer.

If these 'reconstructions' of the history of a victim, innocent or coerced, are to carry more conviction than that which we have been able to adduce, it can only be by demonstrating the likely existence of the person behind the scenes, someone who found in Dawson an ever-active instrument and made use of Teilhard and Woodward as the occasion arose.

This perpetrator, our unknown manipulator, appears as a figure of omnipotence not to be despised by Mephistopheles himself. He has Dawson under the closest surveillance for very many years, as unsuspecting victim for no less perhaps than seven years, alternatively as his unwilling accessory for three years. In two features he is outstanding—his amazingly intimate and detailed knowledge of Dawson's interests and affairs and his complete grasp of the geological, evolutionary, archaeological, and faunistic potentialities of both the Piltdown sites. These qualities have been abundantly illustrated, and his professionalism is well attested also in his ability to obtain and fashion the animal and human fossils and recent bones needed for his great design. His acquaintance with Dawson's movements and inquiries is uncanny. He comes to share Dawson's appreciation of the gravel in great detail, is well aware of Dawson's instruction to the labourers; having conceived his plan, he has unrestricted access to Barkham Manor and makes his 'plant', always certain that his victim will take the bait. His access to the pit is so unquestioned that in his wait of three or four years, while he is preparing the jaw and other fabrications, he can arrange for the rather rare animal teeth and a few other pieces to be safely recovered by Dawson. With an uncanny prescience, he holds back the jaw from him throughout all those years until the climax, the British Museum's participation. He now keeps the excavators well in sight, so that the jaw, the canine, and Piltdown II, three stunning discoveries, are made at the right times and in the right order for establishing *Eoanthropus dawsoni*. He knows, of course, what sort of reception the fossil man has been getting, and will not let the criticisms of Waterston and Gerrit Miller go unanswered. He improves on the archaeological dating and makes archaeological history by providing the bone implement. Perhaps he induces Dawson to put the cranial fragments in bichromate? Anyway, his sinuosity is such that at *every stage from beginning to end* he puts Dawson in flagrant possession of artificially stained and as yet undisclosed fragments—in 1911 before Woodward's participation, in 1915 at Piltdown II, and finally the Barcombe Mills bones. It made no difference at all to the plan and purposes of this perpetrator if Dawson came to a realization of the plot even in 1913. The victim went through a very similar repetition of all his actions of 1912 and before, continuing to dig and write and find more human and animal bones.

This remarkable shadow figure, exerting his complete will over Dawson, was operating before Teilhard de Chardin appeared in the Weald or Woodward went down to Fletching. A tireless and versatile archaeologist and palaeontologist, yet after 1916, or perhaps a little earlier, though the searches went on, the author of the great hoax had lost interest in the whole affair.

---

1. Dawson and Woodward, 1913, p. 117.

2. Dawson, 1913, p .77.

3. Dawson, C., and Woodward, A. S., 1914, 'Supplementary Note on the Discovery of a Paleolithic Skull and Mandible at Piltdown (Sussex)', *Quart. J. Geol.Soc., Lond.*, **70**, pp. 12–93.

4. Le Gros Clark, W. E., contribution in *Bull. British Museum (Natural History)*, 1955, *Geol.*, **2**, No. 6.

5. A report in the *Sussex Express* of 1 January 1954 carries the suggestion that Dawson came into possession of an unusual human skull in 1906:

'Mrs. Florence Padgham, now of Cross-in-Hand, remembers that in 1906, aged thirteen, when living at Victoria Cottage, Nutley, her father gave Charles Dawson a skull, brown with age, no lower jaw bone, and only one tooth in the upper jaw, with a mark resembling a bruise on the forehead. Dawson is supposed to have said, "You'll hear more about this, Mr. Burley." '

6. Mrs. Sam Woodhead, whose husband accompanied Dawson in a renewed search soon after the first cranial fragment came to light in 1908, has a definite recollection of her husband himself finding some pieces on that visit. She believes that these included not only a jaw, but an eye-tooth, which she recognizes cannot be those later reported. We feel that the possibility of a 'telescoping' of those later events with Mr. Sam Woodhead's early search cannot be ruled out as both the spectacular finds are mentioned by her. Dawson also wrote in March 1913 that he and Mr. Woodhead were unsuccessful in their search. As Mr. Woodhead was still associated with the investigation—his analysis of the cranium was reported in 1913—it seems unlikely that Dawson would have made so definite a statement even in a local journal like the *Hastings Naturalist* without Mr. Woodhead coming to know.

Epilogue

✠ ✠ ✠

It remains now only to attempt briefly a final evaluation of the Piltdown affair—its authorship and its scientific significance.

We have seen how strangely difficult it is to dissociate Charles Dawson from the suspicious episodes of the Piltdown history. We have tried to provide exculpatory interpretations of his entanglements in these events. What emerges, however, is that it is not possible to maintain that Dawson could not have been the actual perpetrator; he had the ability, the experience, and, whatever we surmise may have been the motive, he was at all material times in a position to pursue the deception throughout its various phases. For anyone else to have played this complicated role is to raise a veritable Hyde to Dawson's Jekyll. Complementary, also, to the difficulty of excluding Dawson from the authorship—there is nothing that will serve to do this—is the difficulty of accepting his known activities as compatible with a complete unawareness of the real state of affairs.

Yet to condemn Dawson on considerations of this sort is to base the case ultimately on arguments by exclusion. It is true that of the evidence which throws so much suspicion on him, part is derived from his own papers and letters, and that most of the information which has come to us indirectly has not gone uncorroborated, but none of it furnishes the positive and final proof of his responsibility.

So long as the weight of circumstantial evidence is insufficient to prove beyond all reasonable doubt that it was Dawson himself who set the deception going by 'planting' the pieces of brain-case, our verdict as to the authorship must rest on suspicion and not proof. In the circumstances, can we withhold from Dawson the one alternative possibility, remote though it seems, but which we cannot altogether disprove: that he might, after all, have been implicated in a 'joke', perhaps not even his own, which went too far? Would it not be fairer to one who cannot speak for himself to let it go at that?

The end of Piltdown man is the end of the most troubled chapter in human palaeontology. From the first moment of the introduction of *Eoanthropus dawsoni* to the scientific world, the complexities and contradictions of the 'enigma', as Keith continued to call him, took up quite unduly and unnecessarily the energies of students of Man's evolution. This ill-begotten form of primitive man in the several hundred papers devoted to him received nearly as much attention as all the legitimate specimens in the fossil record put together.

The removal of *Eoanthropus dawsoni* does nothing to weaken that record. It provides a clearer picture of the succession of fossil forms in Man's genealogy. When Darwin wrote *The Descent of Man* he had available hardly a single fossil on which to base his arguments, and he relied on a mass of anatomical, physiological, and embryological evidence to illuminate his extraordinarily skilful comparisons between man and other living animals. For Darwin and Huxley the links were still missing, but for us the discoveries of the last thirty years have gone far to provide the fossil fulfilments of Darwin's predictions; but amongst these there is no place for anything like a Piltdown man. Though today we are still far from an understanding of many matters concerning Man's origins, we are in no doubt about the reality of the transformation which has brought Man from a simian status to his *sapiens* form and capability.

Afterword: Piltdown 2003

Chris Stringer

¤ ¤ ¤

Nearly a century after it appeared on the scientific stage, and half a century after it was exposed to hoots of derision, the Piltdown Man hoax continues to fascinate as an unsolved whodunit. A quick search on the Worldwide Web reveals a dozen primary websites on Piltdown and numerous secondary ones, and several of these have been valuable resources in writing this update of the saga. Although many of the sources are creationist ones, exulting in the way in which 'evolutionists' were so easily misled by an 'amateurish' forgery, others present more serious analyses of the historical background, possible motives, or rationale, and the evidence for or against the members of an ever-growing list of suspects behind the debacle. Indeed, the joke that the only participant shown in the fading Piltdown photographic archives *not* to have been named as the forger is 'Chipper' the goose is now close to the truth. When Joe Weiner wrote his entertaining and inform-ative account of the forgery, some of the original protagonists were still alive, but 50 years on, when even those who exposed it have gone, it is remarkable that investigators can still find new angles to the story by revisiting existing sources or even finding new evidence.

Charles Dawson, the amateur antiquarian at the centre of

the Piltdown discoveries, was clearly Weiner's prime candidate
for the forger, although in the Epilogue of his book he drew
back from an affirmation of guilt. As Weiner explained,
Dawson was the first person to seriously search for, and report,
fossils from the Piltdown quarry. In 1912, he and Arthur
Smith Woodward discovered the cranial fragments, and Dawson himself found the mandible. Dawson was present when all
the main finds were made at Piltdown I, including the canine
and the 'bone implement', and he is the only individual
who can definitely be associated with the final 'discoveries' at
Piltdown II. Subsequent to Dawson's final illness and death,
no further significant discoveries were made at either Piltdown
location. Given Dawson's centrality to the story, why should
any other suspects be entertained beyond the most obvious one?

There are various reasons why other individuals have fallen
under suspicion. For the author Ronald Millar,[1] Dawson was
too obvious a culprit and would have known he would be the
main suspect in the event of exposure. Given his profession
and his ambitions, such as election to The Royal Society, he
had too much to lose, it is argued. Millar, like several other
investigators, also believed that Dawson simply did not have
the anatomical knowledge to create a forgery that deceived
some of the best scientific minds of the time. Thus for several
investigators whose ideas I will discuss below, a covert expert
must have secretly collaborated with Dawson in producing
the faked fossils; others have preferred to see Dawson as the
gullible victim of the scheming of others. Let us now consider a
selection of these alternative scenarios in approximately the
order in which they have been advanced.

In the same year that Weiner's book was published,
'Francis Vere' (a pseudonym for the lawyer Francis Bannister)

wrote a very different account.[2] Vere defended Dawson and argued that someone else who was present at all the main excavations was responsible—Venus Hargreaves, the labourer who did most of the real digging at the original site. Hargreaves certainly had numerous opportunities to plant the 'fossils', and if participants such as Dawson and Woodward were not involved, he becomes a prime suspect by elimination. His lack of scientific knowledge and opportunities to obtain the relevant materials, however, certainly require that someone else was passing him the finds to be planted. Thus in 1972 the Belgian academic Guy van Esbroeck suggested that William Butterfield, curator at the Hastings Museum, was Hargreaves's professional accomplice,[3] motivated by a desire to seek revenge on Dawson, who had supposedly misappropriated some dinosaur fossils and sent them to Woodward at the British Museum (Natural History) (now The Natural History Museum, London).

While also defending Dawson's innocence, Ronald Millar[1] argued that the culprit was Sir Grafton Elliot Smith, the distinguished anatomist and human palaeontologist. The scientific theories of both Smith and Sir Arthur Keith (see below) were supported by aspects of the morphology of 'Piltdown Man', but while Keith criticized Woodward's suspect first cranial reconstruction Smith was, according to Millar, suspiciously quiet. Moreover, given his anatomical knowledge, Smith should have recognized the morphological impossibilities of Piltdown at an early stage. Although he was working in Nubia during most of the discoveries, Smith did travel back to England regularly and took part in some of the excavations. But it would have been impossible for him to have operated the hoax alone.

In a posthumously released tape, James Douglas, Professor of Geology at Oxford, named his distinguished predecessor William Johnson Sollas as the main perpetrator of Piltdown, in league with several others.[4] Sollas certainly had the requisite geological and archaeological background, access to collections, and motive—a strong antipathy to both Woodward and Keith. As a relative outsider to the excavations, however, he certainly would have needed one or more local accomplices to plant the materials, which is probably why names such as those of Dawson and Martin Hinton were added to the scenario. Nonetheless, if the intention was to damage Woodward or Keith, it is difficult to see why Sollas would have remained silent over the real nature of the finds.

Earlier, in 1969, Louis Leakey, the anthropologist, had suggested that the Jesuit palaeontologist Pierre Teilhard de Chardin was involved in the forgery, but it was the American palaeobiologist and author Stephen Jay Gould who was most explicit in proposing that Teilhard de Chardin conspired with Dawson, initially as a joke, which then got out of hand.[5] Teilhard de Chardin, then a student lodging in Sussex, was friendly with Dawson, and was at Piltdown on numerous occasions, including when he himself found the canine tooth. Gould argued that Dawson and Teilhard de Chardin were intending to own up to the hoax, but the outbreak of war and Dawson's illness and death prevented this. Teilhard de Chardin's attitude to Piltdown during his later career was certainly reticent, considering that he made one of the most significant discoveries; he made little of Piltdown in his scientific writings, even when specifically discussing human evolution. It is certainly possible that he knew more about Piltdown than he let on, and when interviewed after the exposure, his memory of events seemed uncertain. Although he was over 70 by then, Gould viewed all this as potentially incriminating behaviour.

In 1981, Lionel Harrison Matthews, a retired zoologist, wrote a series of articles on the Piltdown hoax for *New Scientist* magazine, in which he argued for a sequence of forgeries by different individuals.[6] He proposed that Dawson and Lewis Abbott (see below) produced the first finds, but that Martin Hinton and Teilhard de Chardin, aware of the forgery, planted the later discoveries in order to hoax the hoaxers. Hinton was a young geologist/palaeontologist, and an associate in Smith Woodward's department at the time of the Piltdown excavations. Harrison Matthews, a friend of Hinton, was sure that he had known about the hoax at an early stage. He suggested that Hinton and Teilhard de Chardin manufactured ever more ludicrous forgeries intended to expose the hoax, culminating in the elephant bone tool, shaped like a cricket bat. To their dismay, however, all the finds were treated as genuine! This theory was discussed and developed further by other authors, such as the scientist Keith Stewart Thomson.[7, 8]

Sir Arthur Conan Doyle, the creator of Sherlock Holmes, lived in East Sussex. He was a neighbour of Dawson's and had a keen interest in prehistory, participating in the excavations at Piltdown. The American scientist John Winslow and editor Alfred Meyer[9] were the first to argue that Conan Doyle was the author of the hoax, and this idea was developed further by Robert Anderson[10] and (in public lectures) by Richard Milner, both of the American Museum of Natural History. The evidence against Conan Doyle was circumstantial, but reinforced by a series of supposed parallels between the Piltdown saga and Conan Doyle's *The Lost World*, even down to the claimed resemblance between inscriptions illustrated in the book and the geography of Piltdown. Conan Doyle was a spiritualist who had crossed swords over his beliefs with several scientists, including Sir Ray Lankester, a Director of the British Museum (Natural History). So the possible motive could have been a desire to expose and discredit the scientific establishment over their evolutionary beliefs. Although as a medical doctor and amateur prehistorian Conan Doyle had adequate knowledge to conceive the hoax, there remains doubt that he had the opportunities to carry it out. And as with several other suspects, it is difficult to see what was gained when the truth about Piltdown only emerged after most of the main players were already dead. If Conan Doyle was not involved in, or at least in the know about, Piltdown, the apparent parallels between *The Lost World* and the Piltdown story can only be a series of coincidences.

The literary historian Peter Costello[11] and the Cambridge archaeologist Glyn Daniel[12] argued that Samuel Woodhead, a public analyst, produced the forgery, with Daniel also implicating Woodhead's colleague John Hewitt, a Professor of Chemistry at Queen Mary College London. Costello noted comments about Piltdown by Woodhead's son, Lionel, in a letter written to Kenneth Oakley in 1954. Lionel Woodhead claimed that his father was involved with Dawson in the early discoveries at Piltdown, including that of the jaw. In a subsequent letter to Daniel, Lionel said that his father avoided discussing Piltdown but that his mother remembered Dawson bringing bones to his father, asking how they could be made to look more ancient. When the actual Piltdown remains were discovered, Samuel apparently tested them and discovered they had been tampered with, but when the discovery was announced, he kept silent out of misplaced loyalty to Dawson.

Costello found all of this highly suspicious and concluded that Samuel Woodhead was, in fact, the forger, with all the necessary chemical skills. As a reportedly devout Presbyterian, Woodhead supposedly hoped to discredit the theory of evolution, although his subsequent silence over Piltdown then becomes hard to explain. Daniel expanded on Costello's hypothesis by adding Hewitt as a conspirator, based on the evidence of family friends, who recalled Hewitt recounting that he and a friend (presumably Woodhead) had created the forgery.

Joe Weiner thought Lewis Abbott a potential suspect because of his geological and archaeological knowledge (and friendship with Dawson) before he focused his attention on Dawson himself. Abbott was a jeweller by profession, but also an enthusiastic palaeontologist, with a strong belief in the potential of the high gravels of southern England to produce evidence of the earliest Britons and their Eolithic implements. The American academic Charles Blinderman took Weiner's considerations further to argue that Abbott had all the requisite skills and scientific motivation to produce the forgery,[13] and Abbott also figured as a conspirator in several other scenarios. Caroline Grigson, a curator at the Royal College of Surgeons in London, managed to come up with yet another candidate for the Piltdown hoax—Frank Barlow, a preparator and casting specialist at the British Museum (Natural History), who appears in the famous illustration of a discussion of the Piltdown remains.[14] Barlow had exclusive rights on the production of plaster casts of the Piltdown remains and Grigson argued that Barlow had financial motives to join with Dawson in producing the forgery.

Ian Langham, a historian of science at the University of

Sydney, spent many years reviewing the Piltdown saga, and had earlier considered that Grafton Elliot Smith and Arthur Smith Woodward may have been involved. But before he died in 1984, he had privately concluded that the eminent anatomist Sir Arthur Keith of the Royal College of Surgeons was in league with Dawson in producing the hoax. Frank Spencer, an anthropologist at the City University of New York, was appointed to continue and publish Langham's research, completing this in 1990.[15] Their arguments centred on several incidents where Keith, a relative outsider to the discoveries at Piltdown, appeared to know much more about events there than would have been expected without considerable inside knowledge, or even foreknowledge. Langham and Spencer also considered that Keith had concealed his connections with Dawson and behaved suspiciously in destroying their correspondence. As one of the few survivors of the Piltdown saga when the hoax was exposed in 1953, Keith was interviewed several times and was, according to Langham and Spencer, suspiciously vague about events—but he was nearly 90 by then. The Langham–Spencer hypothesis was further developed by the South African anatomist Phillip Tobias in 1992.[16]

The American anthropologist Gerrell Drawhorn proposed another central character in the Piltdown story as forger in league with Dawson: Sir Arthur Smith Woodward.[17] As Keeper of Geology in the British Museum (Natural History), Woodward potentially had sources for all the materials recovered from the sites of Piltdown I and II, and earlier unpublished suspicions were raised about him by the Cambridge anthropologist Jack Trevor. According to Drawhorn, the kudos of studying and publishing such important finds would have furthered his ambition (never achieved) of becoming Director of the BM(NH), as well as confirming the evolutionary theories that he advocated. Drawhorn also considered Woodward's subsequent behaviour suspicious because he failed to carry out potentially revealing studies of the materials and purportedly prevented others from doing so. Further suspicions were raised by his vagueness about details of the Piltdown II discoveries. On the other hand, unless he was very carefully covering his tracks, Woodward continued excavations at Piltdown long after Dawson's death, moving near the site to continue the work after his retirement, and never found anything else of significance.

Henry Gee,[18] an editor at the journal *Nature*, outlined the case being made by the palaeobiologist Brian Gardiner[19] against Martin Hinton as the sole perpetrator of Piltdown, rather than as one of several conspirators as Harrison Matthews and others suggested. In the mid-1970s, an old canvas travelling trunk with Hinton's initials on it was found when loft space was being cleared at The Natural History Museum. Among the items unpacked by the curator of fossil mammals, Andrew Currant, were mammal teeth and bones stained and carved in the manner of the Piltdown fossils. Separately, further items including stained human teeth were forwarded to Currant and Gardiner by Hinton's scientific executor, the palaeontologist Robert Savage. Hinton had long been interested in the geological processes that stained fossils, and Gardiner argues that the staining procedures in Hinton's materials were the same as those used in the Piltdown assemblages. The motive might have been revenge over a quarrel about departmental payments due to Hinton, or perhaps Hinton, like several others, had taken a personal dislike to Woodward.

Shortly after the exposure of the forgery, Martin Hinton indicated in conversations, interviews, and correspondence that he had long had suspicions about Piltdown and knew who was behind it. Hinton certainly had the geological knowledge and access to materials to produce the forgeries, whether in league with Dawson or not (Gardiner believes that Hinton duped both Dawson and Woodward). Before concluding the case against Hinton, however, we should await the publication of new chemical analyses comparing his materials with the Piltdown originals[19] to see how similar were the staining methods. In addition, there could be more innocent explanations for Hinton's materials. His long-term interests in geological staining may have led him to conduct his own experiments on this, perhaps extending to attempts at duplicating the methods used at Piltdown, once his suspicions were aroused.

As the above account shows, several ingenious new hypotheses about the identity of the Piltdown forger or forgers have been produced since Joe Weiner published his first edition of this book. Taken in isolation, several of them seem very credible, but are any more compelling than Weiner's final concentration on, if not affirmation of, Dawson as the hoaxer? As the historian John Evangelist Walsh discussed, there are now several additional reasons to suspect that Dawson was not merely the innocent victim of the malice or trickery of others.[20] Dawson made something of a career of discovering 'firsts', or critical finds. As Weiner reported, in 1893 he had visited the British Museum with the 'Beauport statuette', a small figurine claimed to be of Roman age and apparently the oldest evidence of cast-iron manufacture in Europe. There is now even more suspicion about the authenticity of this object, and Dawson's vague and somewhat contradictory accounts of its discovery, because contemporary first-hand accounts of the excavations strangely do not mention such a potentially important find, and it seems likely to be a much more recent forgery.

Another of Dawson's supposedly 'critical' finds from the Roman period, part of a tile reportedly from the Roman fort at Pevensey in East Sussex, was circulated by him in 1907, but apparently found 5 years earlier. The tile carried an inscription, the first known from Britain referring to the Emperor Honorius, and was a more complete version of one found during systematic excavations at Pevensey, also in 1907. In 1972, both were tested using thermoluminescence dating and were shown to have been manufactured, by the same hand, less than a hundred years earlier, with Dawson the obvious suspect. Walsh further discussed the apparent forgery of an ancient map of the village of Maresfield, near Piltdown. Dawson claimed to have copied his published version from an original drawn in 1724, but the original has never been traced and it was described in 1974 as 'wholly fictitious'.[21] And as well as the examples of which Weiner was aware, further cases have emerged where Dawson either used the written work of others with minimal credit or, in the opinion of some experts, he carried out gross plagiarism.

In my opinion there is now even more evidence to support Weiner's suspicions that Dawson was heavily, and perhaps solely, implicated in the Piltdown hoax. He was the only figure present throughout the main events, and the bizarre 'discoveries' at Piltdown II can only be laid at his door. Although radiocarbon dating did not establish that any cranial parts were in common between Piltdown I and II,[22] my comparison of the molar teeth from both sites supported the supposition that they are from the same mandible.[15] Hence, whoever produced the Piltdown II finds had access to the rest of the jaw that appeared at Piltdown I, and had treated the new find in identical fashion. Only the most convoluted of scenarios could possibly exculpate Dawson from responsibility for Piltdown II, even if such might be possible for aspects of Piltdown I.

For me, the mysteries that remain are whether Dawson had the knowledge and materials to have acted alone, and whether others were involved independently of him, and here perhaps the name of Martin Hinton, at least, stays in the frame for further investigation. In the future we may see the emergence of new documentary evidence, new chemical analyses to track the origin and treatment of the fraudulent fossils, and perhaps even DNA testing to establish the real sources of the '*Eoanthropus*' material. But one thing is certain: international interest in the forgery and forger(s) will continue during the coming years, and Joe Weiner's book will remain one of the key sources of information about this remarkable story.

Websites and references

As I mention above, there are many websites dealing with Piltdown, and at the time of writing the three linked sites listed below provide the most comprehensive coverage. Spencer's two 1990 books contain the best archival sources, while Walsh's 1996 book is, at present, the most up-to-date published review of the saga.

http://www.talkorigins.org/faqs/piltdown.html
http://home.tiac.net/~cri/piltdown/piltdown.html
http://home.tiac.net/~cri/piltdown/piltref.html

1. Millar, R. (1972). *The Piltdown Men*. Gollancz, London.

2. Vere, F. (pseudonym, real name Francis Bannister) (1955). *The Piltdown Fantasy*. Cassell, London.

3. van Esbroeck, G. (1972). *Pleine Lumière sur l'Imposture de Piltdown*. Éditions du Cèdre, Paris.

4. Halstead, L. B. (1978). New light on the Piltdown hoax? *Nature*, **276**, 11–13.

5. Gould, S. J. (1979). Piltdown revisited. *Natural History*, **88**, 86–97; Gould, S. J. (1980). The Piltdown conspiracy. *Natural History*, **89**, 8–28.

6. Matthews, L. Harrison (1981). Piltdown Man: the missing links. *New Scientist*, **90**, 280–2, 376, 450, 515–16, 578–9, 647–8, 710–11, 785, 861–2; **91**, 26–8.

7. Thomson, K. (1991). Piltdown Man: the great English mystery story. *American Scientist*, **79**, 194–201.

8. Broad, W. and Wade, N. (1982). *Betrayers of the Truth*. Simon & Schuster, New York.

9. Winslow, J. and Meyer, A. (1983). The perpetrator at Piltdown. *Science 83*, September, 32–43.

10. Anderson, R. (1996). The case of the missing link. *Pacific Discovery*, Spring, 15–20, 32–3.

11. Costello, P. (1985). The Piltdown hoax reconsidered. *Antiquity*, **59**, 167–73.

12. Daniel, G. (1986). Piltdown and Professor Hewitt. *Antiquity*, **60**, 6, 59–60.

13. Blinderman, C. (1986). *The Piltdown Inquest*. Prometheus Books, Buffalo.

14. Grigson, C. (1990). Missing links in the Piltdown fraud. *New Scientist*, **125**, 55–8.

15. Spencer, F. (1990*a*). *Piltdown: A Scientific Forgery*. Natural History Museum, London/Oxford University Press, New York; Spencer, F. (1990*b*). *The Piltdown Papers: 1908–1955*. Natural History Museum, London/Oxford University Press, New York.

16. Tobias, P. (1992). Piltdown: an appraisal of the case against Sir Arthur Keith. *Current Anthropology*, **33**, 243–60, 277–93.

17. Drawhorn, G. (1994). Piltdown: evidence for Smith Woodward's Complicity. *American Journal of Physical Anthropology*, **95** (Suppl. 18, Annual Meeting Issue), 82.

18. Gee, H. (1996). Box of bones 'clinches' identity of Piltdown palaeontology hoaxer. *Nature*, **381**, 261–2.

19. Gardiner, B. (in press). The Piltdown hoax: a review of the evidence. *Proceedings of the Linnean Society of London*, **00**, 00–000.

20. Walsh, J. (1996). *Unraveling Piltdown: the science fraud of the century and its solution*. Random House, New York.

21. Andrews, P. (1974), A fictitious purported historical map. *Sussex Archaeological Collections*, **112**, 165–7.

22. Spencer, F. and Stringer, C. (1989). Radiocarbon dates from the Oxford AMS system: Piltdown. *Archaeometry*, **31**, 210.

Index

¤   ¤   ¤

Popular Science from Oxford

Galileo's Finger: The Ten Great Ideas of Science
Peter Atkins

Ten great ideas, ranging from natural selection through quantum theory to curved spacetime, are introduced with brilliant imagery in this best-selling introduction to modern scientific concepts. Never before have these core ideas of modern civilization been presented in so engaging a manner.

'This book is one of the best panoramic views of nature's extraordinary symmetry, subtlety and mystery currently on offer'
John Cornwell, *Sunday Times*

The Emperor's New Mind: Concerning Computers, Minds, and the Laws of Physics
Roger Penrose

A fascinating roller-coaster ride through the basic principles of physics, cosmology, mathematics, and philosophy, to show that human thinking can never be emulated by a machine.

'Perhaps the most engaging and creative tour of modern physics that has ever been written'

Sunday Times

OXFORD

Popular Science from Oxford

Fabulous Science: Fact and Fiction in the History of Scientific Discovery
John Waller

The great biologist Louis Pasteur suppressed data that didn't support the case he was making. Einstein's theory of general relativity was only 'confirmed' in 1919 by an eminent British scientist who massaged his figures. Gregor Mendel never grasped the fundamental principles of 'Mendelian' genetics. Often startling, always enthralling, *Fabulous Science* reveals the truth behind many myths in the history of science.

'Everyone with an interest in science should read this book.'
Focus

Eurekas and Euphorias: The Oxford Book of Scientific Anecdotes
Walter Gratzer

Around 200 anecdotes brilliantly illustrate scientists in all their varieties: the obsessive and the dilettantish, the genial, the envious, the preternaturally brilliant and the slow-witted who sometimes see further in the end, the open-minded and the intolerant, recluses and arrivistes. Told with wit and relish by Walter Gratzer, here are stories to delight, astonish, instruct, and entertain scientist and non-scientist alike.

'There is astonishment and delight on every page . . . a banquet of epiphanies, a reference book which is also a work of art.'
Oliver Sacks, *Nature*